House Keys

House Keys

The Essential Homeowner's Guide to Saving Money, Time, and Your Sanity Building, Buying, Selling, and Maintaining a Home

Lisa Turner

ISBN-13: 9781546350316
ISBN-10: 1546350314
Library of Congress Control Number: 2017907020
CreateSpace Independent Publishing Platform
North Charleston, South Carolina

Dedication

To Jerry

Acknowledgements

Thanks to *The Clay County Progress* newspaper staff and Becky Long, Editor, for cheerleading my home improvement columns and providing the encouragement to assemble this book. Thanks to my husband, Jerry, for inspiration and advice, and to Pat Nardy, home inspector and mentor, for her patient coaching. Last and best: thank you *Clay County Progress* readers. Your flow of ideas and appreciation keep me charged.

Preface

The articles in this book were written from 2010 to 2017 for the readers of the *Clay County Progress.* They reflect the mountain and lake ecology of Hayesville, North Carolina, a picturesque town rich in history, populated with wonderful and thoughtful people.

Writing about homes and property brings me great enjoyment. When inspecting, I love using checklists and solving problems. Every time I see a problem, I think of a way a homeowner could prevent it. As I write, I hope to inspire you to not only head off potential catastrophes, but also make you smile. I've tried to weave information and fun together so that you will remember the advice.

Time is short; happiness is precious. These tips and tricks aim to increase your happiness and satisfaction with your time at home, which should not be filled with endless chores. Using the checklists and techniques will bring added value to your home and help you untangle the stress mess.

Your enjoyment of this book will be my reward in writing it.

Table of Contents

1

All About Home Inspections

Getting the Most Out of Your Home Inspection

You've located the home of your dreams. Your real estate agent has drawn up the offer and the offer has been accepted. Now you have a few weeks in which to perform your due diligence – your research on the home. Is this home everything you think it is? When your agent asks you if you want to get a home inspection, your answer should be "Yes." Following the six tips here will maximize your home inspection results.

Get the best inspector. A really good inspector will catch just about every problem that exists in a home, and be able to put the issues into perspective for you so that you can make a decision. Your inspector should be licensed and/or certified at what they do, and have a good experience level. Your agent will give you several names; either call each one, or research them on the internet. The best inspector and the worst inspector will likely charge you the same amount of money, so why not maximize your return?

Be there for the inspection. 85% of home buyers are not present during the inspection, thinking that they will disturb the work of the inspector. Not so. The inspector will welcome you for the inspection, and be glad that they can point out the most important issues to you so that you fully understand what he/she has found.

Ask the right questions. Your inspector is making value judgments as they inspect – would they buy it? Ask your inspector to be honest with you – take advantage of this special "inspector-client privilege" relationship to ask him/her what their concerns are. These are things they won't write in the report, but they will discuss with you.

Ask for a "show and tell" on any items you do not understand. If you are not at the home during the inspection, arrange for a call with your inspector after the inspection to discuss these items. The inspector will tell you things over the phone that may not be in the report.

Maintenance advice. Although not part of the formal inspection, most good inspectors will tell you what items in the home require regular maintenance and will tell you what that maintenance consists of. Take advantage of the inspector's expertise.

Ask for referrals. Who does the inspector recommend for repairs? This is also something you will not see in the formal report, but the inspector will have a few very talented tradespeople that they use and you should find out who these people are.

Follow these tips to get the most out of your due diligence. Buying a home is one of the largest investments you will make; don't put yourself in the position of finding serious problems after the sale and being trapped. This up-front effort will ensure you know exactly what you are getting.

How To Use a Home Inspection Report To Save Money

Should you use a home inspection report to make a home buying decision? Definitely. Realize that I am biased in favor of this advice, because I was once a home inspector. The stuff I found on inspections was a constant source of surprise to me. All homes have defects. While most are minor, all it takes is one big one to blow the budget over to the next county.

I am also an airplane nut. This summer at an airshow, I held a talk about building your own airplane. Afterwards someone came up to me and said they were going to purchase a used homebuilt aircraft that very day. I asked them who they were having inspect the craft before the purchase. They said that they were skipping the inspection, saying, "It flew in here to the show so it must be fine."

If you're buying a house, this is like saying the home's roof doesn't leak water so everything must be fine inside.

Whether you are buying an airplane or a home, get an inspection. The price you pay will pale in comparison to what you might end up spending to correct major defects. Here's how to use an inspection report to save you money.

Don't worry about the small stuff. A stuck window can be fixed. A leaky faucet can be repaired. Missing attic insulation can be added. You can even ask the seller to make repairs. They don't have to, but many times sellers will bend over backwards to make you happy. Skip over the small stuff in the report and go to the section called Major Items. These are at the top of the list in the report and/or in the summary.

The seller is supposed to tell you about significant defects. This rarely happens. This is not necessarily because the seller is covering it up, but because they don't know about it. I have gone into dozens of basements to find furnaces that were recalled because of carbon monoxide hazards and the owner had no idea (and was lucky).

A note about inspectors. Inspectors are picky, especially new ones. Real estate agents know this and it drives them crazy. Agents want you to be happy with your new home as much as you do, and they have to wade through the small stuff too. As long as you pick an experienced and unbiased inspector, pay attention to the major items and if you find something concerning, get an opinion on the cost to correct the fault.

The significant items that you want to focus on are the things that will either hit your wallet in a big way, or will be a risk to your safety and well-being in the home. The maintenance and repair items are handy to have, but every home has an assortment of them.

A few final points. The first is that there is no requirement to do anything with a home inspection report. The report contains the advice of a qualified and experienced inspector so that the seller and buyer can determine the condition of the home and head off unanticipated surprises. The second point is to rely on your real estate agent for perspective. They have done this many times, and as a professional will nearly always give you accurate and thoughtful advice.

Picky Home Inspection?

I recommend a home inspection for sellers as well as buyers. By obtaining a home inspection, sellers can find out ahead of time what the inspector will identify as a problem. Buyers can find out if there are any money traps in the home and get a good idea of a home's general condition.

Readers have asked me some thorny questions over the years about "picky" home inspections, and how to differentiate the little things from the big things. I'll try to answer that question here.

The point of a home inspection is to find the big things that would introduce financial hardship for the buyer that they would otherwise not know about. This includes structural defects, plumbing errors, architectural or builder errors, electrical defects, or lack of maintenance that has compromised any of the major systems in the home.

Home inspections are not specifically designed to catch code violations, although an inspection should point out safety deficiencies. An example is an older home where a bedroom has been added where there is no window in the room. An inspector will flag this condition not specifically because it violates the code for a "sleeping room," but because it is a safety issue. In the event of a fire in the hall, where will the person sleeping in this room escape to? So, "no window" could be a serious problem if the buyers were to assume this room was a bedroom.

Other safety issues that are minor may be called out on the report but are usually not loaded into the "significant" category. These issues are, for example, the spacing between pickets on a deck rail. If the home was built prior to the code going into effect, then there is no violation. While the inspector can list this type of item in his/her report, there is no obligation on anyone's part to correct them. "Picky" report items are usually those items that do not have relevance given the date the home was constructed and do not pose a substantial safety or structural hazard. If you encounter items like this on the report, you should clarify their significance with the inspector.

As long as you are willing to push the inspector a little on identifying the major issues clearly in the report, then the pickier the inspection the better. As a seller you will have a good idea of what will turn up on the next inspection, and you'll have a checklist of items to correct. As a buyer, you'll need to wade through the little stuff to get to the major items, but a picky inspection will not be a problem as long as you get clear explanations.

A final note on pickiness. If the inspector is inexperienced (less than 2 years inspecting), they will be very picky because they don't want to miss anything. This could be to your advantage. If the inspector has been inspecting for a long time, then they may not be able to differentiate what is major and what is not. As long as they capture the "big stuff," pickiness should not be a drawback.

Can a Home Inspection Help You Sell Your Home?

Are you trying to sell your home? Buying and selling has picked up here in the mountains, so the difficulty of selling is going down. This doesn't mean that selling your home is easy. It's not. And it can be stressful at every stage.

Will getting a home inspection help you sell your house? Of course I'm biased, as a former home inspector. But I know that sellers can obtain a big sales advantage by having a pre-sale inspection conducted. Here are the pros and cons.

Getting a seller's home inspection is a good idea if:

- You want to avoid the last-minute surprises of a buyer's home inspection. Sometimes both parties are very surprised at what is found in an inspection. Suddenly there is work, aggravation, and a small measure of irrationality thrown in as both parties wrestle with who will do what. Sometimes unanticipated discoveries will sink the deal. If you have the chance to address these beforehand, you are more likely to keep the sale.
- If you want an extra edge in marketing the home. Any extra effort or component to make the home more appealing to buyers will help set your home apart from everyone else. Getting an inspection indicates to the potential buyers that you are serious about taking care of the home, and selling it with all issues disclosed.
- The home is already in good condition. A home inspection will probably turn up some items that you are unaware of. They will be small items and you can take care of them relatively easily. The inspection report then becomes a marketing tool.

On the other hand, getting a seller's home inspection may not be a good idea if:

- You know that there are numerous problems with the home and would rather not dig up the details on them. This is a little ostrich-like, but some sellers feel that it is better to not disclose defects and rely on the "buyer beware" doctrine even though they know there may be ramifications at contract due diligence time.
- The home has very serious known defects and is being sold as is. There is little point in illustrating the multiple deficiencies

of a home as a seller in an inspection report when you know the buyer will probably be ordering their own inspection.
- If you are asking a high price for the home and sticking to it. You know the buyer will be getting an inspection but you have already decided not to negotiate. Although risky, this tactic may still bring a buyer as the market picks up but extends the time that the home is on the market.

Most home inspectors offer substantial discounts for seller inspections, and it is typical for the inspection to also include a return follow-up with an updated report. Getting a seller's home inspection can give you a substantial advantage in selling your home for the best price, and can reduce the amount of stress that selling a home induces. You will have all of the knowledge without any of the surprises.

Too Few Inspections: Recipe for Catastrophe

With real estate markets picking up, many people have decided to build the home of their dreams. From a small cabin to a large timber frame home, getting the construction details right is critical. As long as you pick a top-quality builder, everything will be fine, right?

Wrong. Even building your dream home from a purchased set of plans will bring errors and omissions. This is the nature of home building. One homeowner I talked to recently said, "It doesn't seem to matter how many homes you build; you think that each one is going to be perfect, and then you discover after you move in that you or someone else forgot something." This is especially true when building the dream home you drew out on a napkin one night over dinner.

If you are building, or planning to build, a home from plans or a custom home that has never taken shape before, I have some advice that will save you some heartache. You will have enough surprises even with the best planning.

Get your own inspector. Hiring a home inspector or qualified independent project GC to look at your home at each stage of construction will insure that you get the best possible results.

Why would you want to spend this money? What could possibly go wrong?

- On one inspection, I discovered that the electrician had drilled six 2 inch holes through two major roof trusses (a big no-no). Result if not corrected: potential roof/gable collapse.
- Installers forgot to flash 4 of the rear windows. Result if not corrected: massive water leaks into the home (tough to correct after siding is on).
- I inspected a spec home for sale where the outside deck was not supported properly. I flagged it as unsafe. Before corrections were made a windstorm caused the entire structure to collapse.
- On one new home the contractor forgot to install a drainage system before backfilling the basement walls. Result: a 2-foot-deep swimming pool in the basement.

This is a fraction of what I used to see when I was inspecting. This is not to cast blame on contractors. The majority of builders are highly conscientious, and our municipal inspectors are excellent. However, there is only so much you can manage without having another fresh and educated set of eyes to catch problems. A home is a complicated assembly of systems.

If you go this route, hire your inspector early in the process. When you contract with your builder, tell them that you have an independent inspector and that they will not get in the way. The person you hire will be your advocate, communicating with you so that you can communicate corrections to the builder in time. Your builder may not like this, because they will think you are second guessing them. But the home is yours, and you want and deserve the best.

Negotiate pricing with your inspector, but expect to pay about $150 per visit. 5-8 inspections are typical. I guarantee that these inspections will end up saving you money on overages and mistakes. There's even a chance you will get into the home and say, "We didn't miss a thing!"

Top Six Myths About Home Inspections

If you have bought or sold a home, you might have experienced an independent home inspection. This type of home inspection is designed to provide both buyers and sellers with critical information about the health of the home's systems – heating and cooling, electrical, plumbing, water tightness, roof condition, and safety. This type of inspection is highly detailed and provides a wealth of information on the home. While this type of inspection is not required, it can help buyers avoid a "money pit" and can help sellers understand what things might turn buyers away.

A Progress reader wrote me last week to say that they recently bought a house and had expected the home inspector to look for termites. After they moved in, they decided to remodel. They discovered that termites had completely eaten the wood structure in 3 walls.

I told them that one of the things home inspectors do not do is inspect for pests, since they are not qualified to identify them. Pest control professionals are qualified to find pest infestations, and should be called in before the purchase. Most of the time your real estate agent will suggest what inspections you should be getting to protect yourself.

This got me thinking about home inspection myths. Here are the top 6 myths you should know about.

- Home inspectors inspect for termites. Myth! Unfortunately for the couple above who believed this, repairs were very expensive.

- You should not attend the inspection on the home you are buying, because it will disturb the inspector. Myth! Inspectors appreciate their clients attending the inspection and know they can fully communicate the issues with them. Sometimes written reports do not explain everything fully. If the clients are out of town and cannot attend the inspection, they should hold a conference call to discuss report items as soon as practical after the report is completed.
- The seller is responsible for fixing everything the inspector finds wrong. Myth! Repairs, even serious ones, are negotiable. The sellers may be able to back out of a deal, however, if the inspector discovers serious defects.
- New construction requires an independent home inspection to get the Certificate of Occupancy. Myth! New construction does require progressive inspections by the municipal building inspector for safety and code enforcement. If you are moving into a newly constructed home, I personally would recommend an independent home inspection also, as it will catch many loose ends.
- If the home's appraisal is excellent, there can't be anything wrong with the home and you don't need another inspection. Myth! A home's appraisal is based on many factors, including market conditions, location, and materials (HardiePlank and granite counter-tops, for example) but does not inspect for systems actually working or structural integrity.
- A home inspection will take about 30 minutes. Myth! A thorough home inspection should take from 2-5 hours depending upon the size and complexity of the home. There are hundreds of inspection points on a home inspection, including walking the roof and crawling the crawlspace.

Now that you are the home inspection expert, you can try these questions on your friends and see how they do.

The Top Six Problems that Home Inspectors Find

Progress readers routinely ask me what the most common problems are that home inspectors find. If you are buying a home, these issues may cause problems for the sale. If you are selling a home, review these to minimize surprises. You may want to hire a home inspector prior to listing your home if you are aware of any of these issues with your home.

- Outside grading. Over 60% of the homes I inspected did not have a steep enough slope, or grade, to lead water away in a heavy rain. If you wonder why you have a damp basement or crawlspace – this could be part of the reason. The first 10 feet out from your foundation should drop at least six inches. This may not be easy to do here in the mountains where we build on the side of a hill. What to do: install a drainage ditch on the up slope to lead water away from your home.
- Missing, improperly placed, or damaged gutters. Closely associated with the grading problem above, gutters that are not functioning will introduce moisture into walls, basements, and crawlspaces. Mildew and mold follow, causing damage to materials, and potentially adding contaminants into the air that you breathe. What to do: Inspect your gutters and downspouts both spring and fall and make repairs. Observe what is happening in a heavy rain – your gutters may not be sized correctly. Go out and look – are the gutters carrying away ALL of the water?
- Heating system problems. This is a serious category with health and safety impact. And it's a large category, covering everything from gas leaks to heat exchanger cracks to wood burning fireplaces that have never been cleaned. I found furnace filters that had not been changed in years, and piles of construction debris still lying inside air supply vents. What to do: Have your heating and cooling system serviced twice a year by a professional. Have a maintenance checklist for the other items (filter schedule for example) and follow it.

- Missing smoke detectors. 65% of the homes I inspected either did not have any working smoke detectors, did not have enough smoke detectors, or the detectors were not in the right places. What to do: check your home now. How many detectors do you have? Are they working? You can push the "test" button on each unit to see if it's working. Buy photoelectric versions if you need more. Install smoke alarms in every bedroom, outside every bedroom, and on every level of your home.
- Undersized or faulty electrical systems. In older homes where small electrical panels (less than 200A) were common, I would routinely see where residents had augmented the outlets with extension cords and multipliers. I would also discover additional distribution panels that had been installed improperly. What to do: Minimize the use of extension cords and multiplex outlets. If you see lots of this in a house you are buying, get an electrician or home inspector to check it out.
- Expired roof materials and cracked plumbing stack gaskets. Asphalt shingle roofs have a life of 15 to 30 years depending upon quality and material. Typically, homeowners move in and forget the roof unless they suffer a leak. What to do: Hire a home inspector or handyperson to inspect your roof every 3 - 5 years.

Consider starting and keeping a maintenance checklist for your home if you are not already doing it. Why court surprises?

For very detailed questions and answers on every possible aspect of home inspections, see *Questions and Answers: Everything You Wanted to Know About Home Inspections* at the end of the book.

2

Home Inspection True Tales

Home Inspection Surprises 1

Readers routinely ask me what odd or unusual things I have encountered during my home inspection career. Although I try to anticipate almost anything on a home inspection, I have definitely been surprised.

Surprises included water leaks in the crawlspace, parts of the house without access (no hatch or entry to the attic or to the crawlspace), and subflooring held up by barstools. In several instances access to the crawlspace was via floorboards that were not fastened down. I felt like Nancy Drew searching for mystery clues.

One winter I inspected a lovely 6,500 square foot manse on a hill in the woods. The home was vacant, and the seller had turned the water back on the day before the inspection. The real estate agent and I arrived in the morning and entered through the front door. We both heard water when we entered the foyer. We walked towards the sound and ended up at the top of the stairs to a very dark basement. I pulled out my flashlight. About halfway down the stairs, inky water lapped at the carpet. An outside faucet had burst, releasing thousands of gallons of water into the walls of the basement, creating an unintended swimming pool.

Crawlspaces are important inspection areas. You can find water leaks, missing ductwork, loose insulation, structural deficiencies, and plumbing problems. I always crawl to every corner to look at everything. One time I entered a 100-year-old home's crawlspace that had about 24 inches of crawl clearance. I pointed my flashlight ahead of me as I squiggled and squirmed over the damp dirt floor from the outside. Out of the corner of my eye I saw quick movement and reflections of multiple eyes. I stopped, frozen in my crawl, thinking I was coming up on a den of snakes. My heart raced. The movement ahead of me stopped. I trained my flashlight slowly back to the corner. At least 3 sets of big eyes and 4 sets of little eyes were looking at me, motionless. Raccoons! Whew, I took a breath. Then I noticed 4 scissors jacks to my right, holding up the floor of the old home. These jacks are the kind you keep in your car to change a flat, not to hold up a house. As I kept the light pointed toward the raccoon family, I noticed with another stab of fear that the jacks were rusted through and were probably on the verge of collapse. If you had been there, you would have seen me doing the strangest of backward wiggles to exit that particular crawlspace.

In another inspection of a vacant home, a full length "secret" cabinet door led to a large closet space with a door at the end. This door was locked but the key was in the knob. I turned the key in the lock, and rotated the knob. The door creaked open towards me and dust particles floated in the beam of my flashlight. Behind the door was . . . nothing. I played my flashlight into the void, discovering a 7 foot drop to a dirt floor with no stairs!

The moral of these stories is: Bring a good flashlight and a ladder with you to inspections, and know how to wiggle backwards.

Home Inspection Surprises 2

I received notes from readers who wanted to hear more home inspection surprises. The work is so varied that plenty of unusual stories spring from the profession. I suppose it is a little bit like doctors, attorneys, and real estate professionals, who always have a wild story to tell.

Home inspector working conditions can be luxurious, as in the inspection of a 12,000-square foot mansion, or dangerous, as in the inspection of a very remote vacant home in a mountain forest. I don't know if you've explored some of the back roads around this part of North Carolina, but many of them lead you to some strange situations.

One day my husband and I were cruising around exploring the back roads. One road became imperceptibly narrower until the truck barely fit between the tall trees lining the lane.

"This happens all the time," said the friendly woman when we arrived in her carport.

"Sorry, we were driving around looking," I said, sheepishly, as I exited the truck and directed my husband out of the carport and in to a small grassy area to get the truck relocated 180 degrees to leave.

"I suppose we should put a turnaround in, or a sign down the road," she said as I got back in the truck to leave.

"Nice meeting you," I said.

"Yeah, bye, have a great day."

When you think about it, where else could you drive into a stranger's carport accidentally and be treated like family?

Back to inspection surprises. One day I was inspecting a large cabin in the Highlands area. The cabin had been vacant for a year and now a buyer was interested. Everything looked fine from the outside and passed my first walk around looking for problems except for one strange item: a PVC tube about two inches in diameter hanging out of the soffit through a vent. Water, or something, dripped slowly from the end.

"What in the world is that?" I said to myself.

The surprise greeted me as I crawled up my ladder to the attic. Directly in front of me was a children's swimming pool about 6 feet in diameter and 18 inches high. I stood up as well as I could under the roof and trained my flashlight on the pool in the darkness.

"What!?" I exclaimed, looking at the inside of the pool. It was full of water!

Then I trained my light on the roof interior. I saw where the roof damage was, and moved over to the back side of the pool, careful to keep my footing on the beams so as not to fall into the room below. Why wasn't the pool overflowing? I thought. Then I saw the PVC tube exiting the pool right at the water level and running across the rafters to the soffit.

Got it! That was the same tube I was puzzled over on my exterior walk. I snapped a picture of the arrangement because I knew no one would believe me.

Although a clever way to keep a roof leak from entering your home, it's not a solution that I recommend to readers.

Home Inspection Surprises 3

When inspecting homes, ordinary doors can provide a surprise. Some doors lead to rooms, some doors lead to a dark void, and some doors are curiously locked. Sometimes you get all three.

I was inspecting a large vacation home north of Cashiers on a fast running creek. It was full of boulders, twists and turns, and waterfalls. The drive to the home was narrow and steep, leading to a heavy gate. The remote the agent gave me worked, and the gates slowly opened on complaining hinges.

The house was beautifully built into the side of the granite ledges, with stunning floor to ceiling windows. Although the home had a small footprint – perhaps 1500 square feet – two stories towered upwards, taking advantage of the very steep lot. The home had been foreclosed on, and was now vacant.

The first part of the inspection on the first floor revealed no anomalies. I started up the stairs to move upwards and noticed a closet door with a deadbolt lock. When you see something like this, owners are usually trying to protect something. Normally I note in the report that I could not access the closet or room, but in this case the bank was the owner and I doubted that they knew anything about this locked door.

I quickly got on the phone to the real estate agent.

"I'll call the bank," she said.

Three minutes later the phone rang.

"No one has a key to that door. If we did I'd say enter and report what you find. Can you pick it?"

"I'm no locksmith. No problem, I'll put it in my report," I said and hung up.

But I was curious.

I ran my hand across the top of the door trim which is where I "hide" a key. My fingers encountered an object with Velcro stuck to the trim. A key! I put the key in the lock and tried rotating it. It worked! Leaving the key in the tumbler, I turned the knob and opened the door.

A black void.

I pulled out my flashlight and aimed it into the area. A black metal circular staircase came into view. Now I felt like Nancy Drew. I started slowly down the narrow stairs and began to hear the sound of water. When I reached the bottom, my feet were on an uneven stone floor and I was in a room about six by six feet with two more doors in the walls. I looked around for a switch. I found it on the opposite wall. I flipped the switch and light filled the room. I was amazed to see that the walls were carved into the cliff.

One closet was a tiny space with an electrical box. The other door was locked with a deadbolt like the one upstairs.

"Oh! I left the key upstairs," I said to myself. "Shoot, I'll have to go back up and get it."

I went back up the circular staircase to retrieve it. I moved back down the stairs to the locked door. The key worked, and I opened the door. I was in a very narrow passageway. The walls were solid rock and

I could see the furrows where blasting caps had been used. I was feeling a little claustrophobic. Should I keep going?

The sound of water grew stronger as I moved slowly down the cavern path. After traveling 12 feet, I was suddenly outside! The waterfall that was visible from inside the home was directly in front of me.

What a surprise! Never underestimate what might be behind a locked door.

Crawlspace Surprises

My career as a home inspector routinely turned up unusual circumstances. I always had to be prepared for the unexpected. Especially when I went into that uninviting area underneath a home called a crawlspace. This name perfectly describes the area. You find creepy crawlies in this space, and you usually have to crawl in this space.

I suit up in head to toe polyethylene coveralls when I crawl into the spaces underneath a home. The tiniest of crawlspaces yield the most clues about the condition of the home but it is also one of the most dangerous places you can enter. From brown recluse spiders to hibernating vermin to a variety of snakes, it must be navigated with care. I wear an airtight mask with breathing filters and a Plexiglas face shield to protect me from a variety of mold, bacteria, and spores. Sometimes I feel like the guy in the spacesuit in the movie "Alien."

One day I was inspecting with another inspector in Cashiers. We arrived at a foreclosed cabin deep in the woods. One of the first things we do is look in the crawl to see what conditions are. As we peered into the darkness, we saw a pool of water and a mold forest hanging down from the floor joists, dripping. We did not like what we saw and considered not going in. But after several hours of inspection inside the house itself, we could not identify the water leak.

I opened the crawlspace door and used my high-powered flashlight to see where I was going. As I got closer to the black pool of water I felt a chill go up my back. I turned around and saw multiple eyes staring down at me from the top of a beam where insulation had fallen down. I banged my light on a post and they scattered. Then the other inspector shouted, "I think it's coming from the bathroom!" Sure enough, as she tested the fixtures, water came gushing down into the space in front of me. I backed up from the black pool and took photos. Then I hightailed it out of there as fast as I could crawl.

Another time I went into a crawlspace and was about as far away from the entry door as I could get, inspecting a beam that was only partially supported by a pier. As I began taking photographs, I heard the crawlspace door bang shut and lock. Oh no! I was trapped! I put my camera away and started shouting as I moved toward the little 28-inch door. "What?" I heard from the outside. The door opened and there was the yard man, just as surprised as I was!

As you can imagine, I do carry a lot of equipment into a crawlspace. A fully charged back up flashlight, a cell phone, and pepper spray for vermin are the most important. A small pry bar I carry probably would have opened the door that the gardener closed. But one thing was for sure: I never knew what the next surprise would contain.

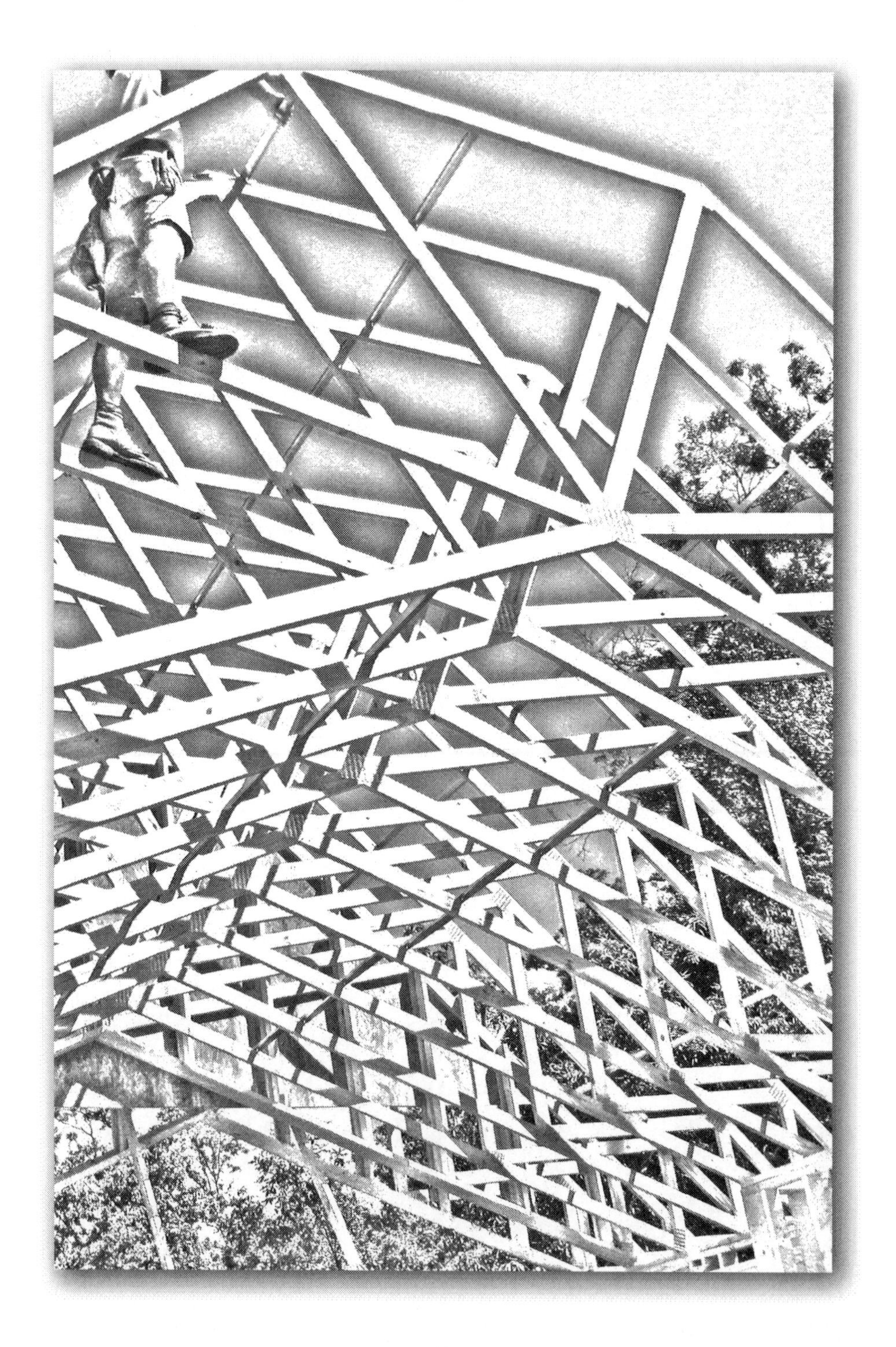

3

Homes: Building, Buying, and Selling

Making Your Home Buying Decision

You've been looking for home to buy for a year. You know what you want. The deals are great. Today you found your dream home. Or, at least you think so. What next?

You think you've made the decision; but actually, you're at the beginning of your buying process. How do you know what you're getting? How do you conduct your "due diligence" on repairs that are necessary? How do you weigh the pros and cons of the repairs versus your love for the house?

Buying a home is a very big decision. You may have spent a lot of time looking at home after home after home . . . your real estate professional has been patient and helpful and now you really think this is the one. You've signed the purchase contract and you've scheduled your home inspection.

Hello. I'm your home inspector! I need to explain what will happen after I give you your home inspection report. Right now, you are basking in that new house glow. You are relieved and nothing can keep you from your dream.

But wait! When you get your home inspection report, you may have an emotional reaction. When you went through the house and said "this is it", you may not have realized that it was not PERFECT. It may have looked PERFECT to you, however, so if you get a list of "problems" in the inspection report, what will you do? You may be surprised; you may be upset, or you may be relieved. I do know that few homes come out of a thorough inspection without something – however small it might be – that needs adjusting, fixing, or replacing.

I wrote this so that you feel better about your inspection and what comes out of the inspection, and how to balance the information you now have about your prospective home with your purchase decision. Large and important decisions should be made with information – and this is what your report will give you.

We also know that emotion plays a role in our decision-making process; sometimes producing what we call irrational decisions. The path that I'm going to lay out here for you will take that into account and help you make the right decision based on what you feel is important.

First, know this about your home – it is not perfect. All houses have some degree of issues, problems, or unusual maintenance characteristics due to the natural complexity of the systems – from plumbing to roofing to electrical to heating and cooling – which work together to make your home livable and enjoyable. Homes are not static, but rather display almost a living quality of needing regular attention to stay well. Vacant homes decline very quickly when this attention and care is lost.

Only 2-3% of homes are what we'd call a "lost cause". These homes have so much wrong with them that it is impractical to spend the money and time to fix everything. If this is the case with a house you are looking at, you will know it before you hire me, and you'll move on. If you find a "fixer – upper" – (8-15% of homes) that is worth investing in, then my report will give you a punch list of repair items to get you started and confirm the extent of the problems you already knew were there.

The rest of the homes – 80 – 85% - are good to excellent. The truth is that most people do take good care of their homes and you will be moving into a home with a solid maintenance record. I'm assuming this is your situation. Let's use this "good house" scenario to demonstrate the decision-making process. Here's how to use your inspection report.

Take a look at the Report Summary. I will tell you about the home in an overview format and highlight any major issues or problems. Only repair items or safety items are put in the summary. Improvement items, comments, observations, and items to monitor will only be in the main body of the report. For example, if I see a stain of some sort under a sink vanity cabinet, and I test it with a moisture meter and it tests dry, and I cannot produce any leaks by operating the faucets and running the water – then I will tell you to "monitor" this area – it looks like there was a leak at one time and that it was fixed – but I cannot be sure unless there are no leaks over time. An improvement example might be where the spacing on the deck pickets is wider than what is used nowadays – it was fine when it was constructed, but when you replace the pickets in the future you should use the newer spacing. It's not a repair, so it does not go in the summary. But it is something you should know, so it does go into the body of the report.

The Summary items are repeated in the main body of the report, so go there to see the photographs and understand the defects.

Use the Summary, or defect list, and determine what kind of contractor or handyperson you will need. Roof – roofer; grading – grader or landscape professional; broken downspout – handyperson; electrical mistake – electrician, and so on. Most of these people will not charge you for a quote. Work with your real estate professional and your home inspector to identify qualified people – we don't get kickbacks! What we care about is finding and referring honest, competent people. In some cases, you may want to get more than one opinion or quote. A good example is HVAC (heating and cooling), and roofing – more than one quote gives you a comfortable range to work in.

You may also want to get quotes for the "Monitor" items. For example, if the water heater is working fine but it's 25 years old – about 14 years over its anticipated lifespan – you will want to know how much to budget for if it breaks suddenly.

Doing this research before your "due diligence" period on the home may be a tight squeeze – the contracts periods are getting shorter and shorter - but it's important unless there is so little wrong that you are not concerned, or the items are clearly a known quantity. If you need to extend the due diligence period to truly ascertain the cost of repairs, then do it – it will not be a pleasant task for your real estate professional – but <u>you</u> are the person purchasing the home. You will be smart to make your decision with all the information that you possibly can.

Now you have your repair quotes as well as your future maintenance budget and you're making your decision. Take a blank sheet of paper and draw a line down the center of it. On the left side top put a plus sign. On the right side put a minus sign.

Now answer the following questions, putting the positives on the left and the negatives on the right. Put a number in each column, from 1 to 10 on the strength of the emotion (or reality) – in any case, how you feel about it. Here is an example: LOCATION – PLUS SIDE - <u>paved roads</u> +8 (how good is it?); NEGATIVE SIDE - <u>long drive to store</u> + 2 (how bad is it?).

Use these categories, but feel free to make up your own!

Location, land or site beauty, house layout, cost of repairs or upgrades, maintenance budget, purchase price, energy efficiency, age of systems, water quality, road quality, internet availability, access in winter, yard maintenance, etc.

Now add up the points on both sides. What do you have? If it's heavily weighted on the plus side in spite of the defects that your inspector has identified – and you have a clear idea of the costs going forward – then your decision is easier and clearer.

The reason I tell you to use your emotional self when assigning points is that your emotional "brain" – your heart – is actually pretty smart and uses logic, amazingly, to make decisions. * Have you ever made a decision that you thought was based completely on "logic", or your brain power – and later found that it should have been made differently? So, take your time – this is, after all, a big decision – and carefully consider all of the factors that you think are important.

I hope that your inspection report and some of these ideas – have helped you make your buying decision easier!

Footnotes:
* Goleman, Daniel, Working with Emotional Intelligence, Bantam Books, 1998. Article 3

Mistakes to Avoid When Building Your Custom Home

Are you thinking of building the home of your dreams? The good news is that you can have everything you want in a custom-built home; the bad news is that common mistakes in planning your home will make it expensive to build and you'll live with errors you wish you'd caught earlier. The custom home I am talking about here is the home you sketched out on a napkin over dessert five years ago. True custom homes are designed and built from scratch, and in some ways are similar to a scratch built airplane; everything has to be reviewed and checked multiple times to make sure it will actually fly.

This process is fun and you will end up with something that no one else has seen before. However, the risks of designing and building a custom home are high because of unknowns. Unknowns translate into money and risk.

Follow these tips to minimize extra expense and surprises.

- Choose your designer or architect carefully. Decide on a flat fee; paying by the hour will get you in trouble because custom home designs must be reviewed many times with many iterations. Make sure your designer uses a program that will allow you to see the plan in 3D. You will be able to virtually walk the home. Are you the designer? Great. Use a 3D design program. Plan on saving a lot of money but investing a lot of your own time.

- Review the final plans again after you think they are perfect. If you or your designer skips this step, you will be making lots of in the field changes, driving up cost.
- Once your perfect plan is done, blow up and print out the layouts for each room and analyze in detail where everything will go – from furniture to light switches. You will have more changes.
- Pick a builder who is willing to make plan adjustments on the fly. Builders hate changes, but it is necessary with a custom home. Even with the best planning, you will discover dimensional problems that need to be corrected as the home is going up.
- Research your builder bids carefully. How much experience do the candidate companies have in building scratch custom homes? Once you have bids based on the plans, you should visit three of the homes each builder is building/has built – one or more during framing, one or more after drywall, and several after the owners are in the home. If you can identify owners who have been in one of these homes for a few years and interview them, you will gain important knowledge about the process and the builder.

Even after all of this planning and careful management, you will see at least one thing you missed when you move in. I don't know why this happens. I expect it is part of the "natural law of home-building."

If all of this sounds scary to you, then a better path will be going online and choosing a ready-made set of building plans. But if this sounds exciting to you, and you have a flexible budget, you will have a one of a kind dream home to move into.

Building Your Next Home: Architect or Designer?

Thinking about or planning to build a new home? Whether it is an 800-square foot cabin or a 6,000-square foot manse, design is critical. The design characteristics of a home drive resale value, cost to heat and cool, cost to build, how many mid-build problems you will have, and the overall functionality of the home.

If you have found a design you like in a plans catalog, all that work has been done for you, but you should still request a 3D plan "walk-through" on a computer before going ahead. 2D plans always look large until you are standing in them!

If you picture your new home in your mind's eye and have not found it anywhere in a plans book, then you should hire an architect or a designer to render your dream with a detailed set of plans for construction.

The difference between an architect and a designer is education, experience, and depth of services. You should choose the one that fits your needs.

An experienced designer is a good choice if you:

- Already know what you want
- Have some home construction savvy
- Don't need a lot of extra services, such as construction stage inspections
- Have the time to get involved in design and construction decisions
- Have a tight budget

A registered architect is a good choice if you:

- Have a general concept of your home in mind but need someone to translate it
- Would rather stay out of the construction details
- Need someone to oversee every stage of construction to ensure the design is carried out correctly

- Do not have a lot of time to get involved in the designs and want to be presented with high level design choices
- Need a full range of building services

Everyone is concerned with construction budgets. You should ask both designers and architects to quote you their fee as a percentage of the construction costs or square footage, with an upper limit that you agree to. A flat fee is ok as long as there is an upper limit. I do not like paying fees by the hour because they add up fast and will surprise you. Designers typically charge between 3-8% of a construction budget, and architects typically charge from 8-15%.

Both designers and architects are competent and professional, but you should choose several and interview them just as you would when choosing a doctor or attorney. You will end up spending a lot of time with this person during the design and construction phases, so a good personality fit is important.

If you want creative flair in your design, you are more likely to get it with an architect. If you want the latest in "green" building technology and small homes, you are more likely to find these with a designer. These are generalizations and you should ask about what is important to you in the interview.

So, the answer to the question "architect or designer" is really about you. If you want creativity combined with original design and a complete services package you will want to hire an architect. If you know what you want, are construction savvy, and don't need a full range of services then a designer can save you a chunk of money.

Understanding Construction Contracts

Often the most difficult part of building a new home is getting the construction contract right. The more due diligence that is put into this stage of your building planning, the less heartache there will be later on when the home is going up. The excitement of building a home can be so great that owners gloss over the contract details, saying, "It will all be fine." But I can tell you from experience that this is not always the case. Here are some key tips to keep your project running smoothly and on budget.

Apples to Apples. As you review construction bids, pull out your notepad and make a matrix of what each builder is offering you and the price. Contractor bids are all different, so this step is not easy. If you take the time to assemble this analysis, it can save you thousands of dollars on your project and shave months off the build time. In general, the best builders will be the most detailed. They know that to leave something vague in the contract or out entirely will mean trouble later.

Make a Wish List and an Assumptions List. You've had the discussion with your chosen contractors of what you want. They are not going to fully understand what you want until you make it specific. If you don't supply the details, they will do what they think you want, and this may not sync with what you really want. Your Wish List is composed of detail around key features, such as the color and type of rock for your fireplace, the type of insulation used in the walls and attic, and what the driveway surface will be. The Assumptions List should include things like adding an ERV (Energy Recovery Ventilation) to the heating and cooling system and a steam humidifier for wintertime comfort. If you ask the HVAC contractors if these are included in a standard installation, they will say no. Yet, an ERV is a must-have for modern, tight homes to bring in fresh air and exhaust stale and polluted air for a healthier environment. Better to find out it's not included now than later! How do you think I found that out?

Fine Tooth Comb. When you think you have everything "nailed down," take another long and critical look at the final builder contract. Find a friend or family member willing to review it also. They will find things that you missed. Check the sections for the things on your own lists, and check their math. If there is a mistake and you sign the contract, guess who ends up paying? That's right – you!

Minimize Mid-Stream Changes. A custom home will never go exactly as planned. This is normal. But you want to catch issues and problems immediately. To do this, you should walk the home after every day of contractor work. If you cannot do this, then hire someone who can. If you catch something going off the rails as soon as it happens, it is always fixable, and it is almost always the builder's responsibility. If it goes unnoticed until later, you will usually have to live with it. When this involves dimensions, door and window locations, and fixtures, it can be disappointing.

Don't Lose Your Shirt on Home Building Contracts

If you have built a home or hired a contractor for large home renovations, you know that the contract details are critical. That flush of surprise when you realize something was not included in the agreement, and you have to borrow money to cover it – is a moment you will not forget.

Here's a quick guide to help you stay out of trouble on contracts and contractor details. A few simple steps can avoid financial disaster and endless workmanship issues once the work is done and the contractor is gone.

Homebuilders and contractors want you to be happy with what they do. It's the foundation of their business. But some contractors are more detailed than others. Misunderstandings spring up quickly when things are not spelled out in detail, and in writing.

- Do your research. Talk to people who have worked with the contractor. Don't just talk to the folks that the contractor wants you to; find others who have used the company and get detail on what they liked and didn't like. Check online ratings and the Better Business Bureau. Make sure that your GC is licensed, registered, and insured.

- Negotiate. There is a lot of opportunity to save money on your home or remodeling job if you'll just politely ask if they can do better on the pricing. It may take several rounds of discussions, but it will be worth it.
- Get a guaranteed maximum price in your contract. Whether fixed price or cost plus, you need to have some control over what you'll end up paying. The more detail in the contract, the better. Actually read everything in the contract, and take issue with what worries you, or get explanations for what you do not understand.
- Ask for detail. The contract should list everything right down to the model numbers and quantities. This protects everyone. Yes, it's more work. In the end, you and your contractor will be glad because it clears up confusion and disagreements. The contract should also detail anything that is not included, that the owner is willing to cover or do, and "allowances" – the cash needed to complete the details that have not been hammered out by contract signing.
- Stay on top of the paperwork. Even though your contractor is managing the labor and materials, you need to be asking for regular updates on where you are on the financial continuum, and make sure that what was contracted for actually happens. An attentive customer makes the contractor manage the details more thoroughly.
- Get a warranty. Most homebuilders cover labor and materials for a year from the date the work was completed. Get these details in writing. Hire a home inspector to inspect the work upon completion, and then again about 10 months later to find anything that went wrong after the work was done. These inspections will pay for themselves.

Don't rush. Planning and not pushing will be appreciated by your contractor and you will end up with a higher quality outcome. Builders hate it when people say, "We need to be in the house by December

15th." Better to be inconvenienced and end up with a home without troubles then to drive your builder nuts on schedule. Be flexible and realize ahead of time that schedule always slips. This is normal.

Follow these tips and then enjoy your creation!

Save Money Building Your Next Home

In an earlier column, I talked about nailing down the details of your construction contract so that the number of surprises would be minimized when you build your next custom home. But problems and changes will come up anyway during the building process. How can you best handle these potentially expensive surprises? Follow these tips to save yourself thousands of dollars.

Bring an expert. You should be inspecting the work yourself or have arranged for someone to do this on your behalf. Unless you are a contractor yourself, bring a friend who understands construction methods at least once a week to your project. Make a list of what they find. Even the most conscientious builders and their contractors are going to miss things. Who ends up paying for mistakes? You do, unless you catch it close to the time it happens and notify your builder. If you forgo these inspections because you are busy, or do not want to pay someone to do this, you are likely to pay another 15% for the home, or live with mistakes that could not be corrected.

Make a list of the items in your builder's contract that say "variable" or "depends on." These are items that are not fixed in a Fixed Cost Contract. These are the surprise areas. Any builder who has started a custom home knows that not everything can be predicted – from weather conditions holding up the schedule and/or damaging work already done, to unintentional errors caused by sub-contractors. Areas that cause over-runs include excavation, footings, and plan errors. Ask your builder to help you calculate what these things could cost worst case and be ready. If things go well, you will be happy.

Double-check the materials list. Are you completely sure you picked out all of your fixtures and appliances? Lighting? Where wall switches go? Review these in detail. If you have gotten so close to the plans and lists that you're not objective, ask someone else to look at it. When you move in to the house and realize that the bathroom fixtures are not the ones you chose, it will not be the builder's responsibility. How do you think I found that out?

Keep a list of action Items for the builder. Your builder will appreciate the attention you are paying on your home; it makes his/her job much easier to have an informed owner. Most top grade successful builders have a dozen or more homes underway at any given time and a limited number of superintendents. They may not get to the project every day, but you will. Keep a list and email routine items to the superintendent for action once a week.

Builders will warranty your home for a year. Many owners simply move in and forget about the warranty unless something obvious goes wrong. This is a mistake. You should keep a detailed list of issues you find – and you will find them – and once a month email them to the builder. This helps the builder by allowing them to schedule your items. Items can include loose toilet tanks, grout cracking, nail pops in drywall, vinyl floors lifting, windows that are stuck, places they forgot to caulk, adjusting the heating and cooling vents for balance, etc.

When you take the time to actively manage your custom home project along with the builder, you will save yourself from the typical financial shocks and end up saving yourself a considerable amount of money. You'll also end up with a higher quality home.

Why You Should Have Your Custom Home Inspected by a Home Inspector

Those of you who have had homes built might recall finding issues with the home after moving in. A builder will tell you that this is completely normal – there is no way that they could have identified and corrected everything in the home, given the complexity of the systems. "It goes with the territory," one builder told me.

Home builders try very hard to build the best house for you and not make mistakes. But builders are busy, and managing subcontractors is not an easy process. The superintendent can't be at your house all the time overseeing the work.

Although these issues can be minor – the cold-water faucet runs hot, a light switch does not work, a window is stuck, a threshold seal is missing, door hardware is loose . . . the problems list can be a glacier where you do not see the big issues hidden below the surface.

The big issues can include ungrounded electrical outlets, plumbing pipe unions installed upside down, heating and air conditioning ductwork hooked up to the wrong zones or cross connected, missing return air ducts, undersized electrical circuits, decks and patios poured with the wrong slope sending water into the house, and incorrect or missing weight bearing structure (as a contractor and former home inspector, I have personally seen all of these).

As experienced and competent as our city building inspectors are, these are not all items that they have on their inspection list. Building inspectors will be paying great attention to code items, which are designed to keep us safe, but might miss patio drainage slope because they are not looking for it.

So . . . what do you do?

You have two choices. The first is to leave it up to the builder and hope that he or she will catch the errors. Or, you can hire a licensed home inspector to perform several inspections during the course of the build. I recommend an inspection prior to drywall, and an inspection just before your walk through with the builder. I also recommend that you hire a structural engineer and have them perform an inspection just before concrete is poured and before drywall.

Both professionals will provide you with a detailed report and list of issues. It's your money and your contract, so the builder should actually be pleased that you provided the punch-out list to him or her. It makes their job easier, since they are doing their best to identify all the issues themselves.

Arranging for and paying for these inspections will not be inexpensive. Remember that the builder already has an engineer in his/her employ for your custom home. What you are doing is spending the money to prevent issues and problems that are sure to cost far more once identified after you move in or when you sell the home. The more eyes that you get on the project to make sure things are correct the better.

Would you rather spend a couple thousand dollars while the home is being built (less than 1 percent of the average custom built home price of $350K - $600K here in the mountains) or wait with fingers crossed that all will be perfect? Based upon what I've discovered in home inspections, the extra money spent will pay dividends.

Tips for the Home Seller Part 1

Selling your home? Oh joy! Isn't this fun? You bought your emotional suit of armor to put on every time your agent calls suggesting you drop your price again, and you know when you finally do sell, the buyers are going to ask for everything else in the home, including your Louvre art collection. So, let's sell your home more quickly so you can get over the suffering.

In my experience as a home inspector, I have seen the issues and problems that home sellers face. Taking a broad perspective on the matter, I assembled 8 tips to help you sell your home faster. These ideas come from experience gleaned from home contractors and renovators, real estate professionals, home stagers and decorators, home systems trades people, home inspectors, and successful sellers. Most of these tips are simple but most sellers do not follow them. If you do follow them, you will be more successful. These are things you can do to help your home sell faster, and get the edge on the competition, without laying out a lot of cash.

Get the Best Agent. Just as you would do your research on contractors to fix your home or find a new doctor, spend some time finding a good real estate professional. Have they been selling homes recently? The current economic climate has seen many real estate offices go out of business, so the ones who are selling now are REALLY good. Interview the top 3 selling agents in your area and then go with the one you can communicate best with. Then follow their advice.

The Front Yard. Some agents call this "curb appeal". It's true: what your home looks like on the outside is a telltale sign of what's on the inside. The outside tells the prospective buyer how detailed you are, how thoughtful you are, and how well maintained your home is. Trim overgrown vegetation. Is there vegetation growing around the windows, on the roof, or on the walls of the home? Trim them back. Mow the lawn more often. Power wash the driveway and paint it. Put down new mulch. Now go outside and stand in the street. How does it look? You'll know what still needs attention.

The Entrance. Power wash and paint. Replace the door hardware. Put down a new welcome mat. It should look like a brand-new home. Keep this area spotless.

Light. When your buyers come in, make sure there is plenty of light. You can't have enough light. All agents know this and will turn on every light in the home while you cringe thinking about your electric bill. They know what they are doing. Replace low wattage bulbs with high wattage bulbs throughout your home. This is not the time to be saving energy.

While these tips have been in the real estate professionals' toolkit for a long time, sometimes home sellers do not realize how critical they really are in drawing and keeping buyer interest.

Tips for the Home Seller Part 2

Are you selling your home? What can you do to get the "edge" on the competition and make your home stand out from the crowd?

These ideas come from experience gleaned from home contractors and renovators, real estate agents, home stagers and decorators, from home systems trades people, home inspectors, and successful sellers. We covered the first four last week: get the best agent, spruce up your front yard and entrance, and turn up the lights. Here are four more things you can do.

Smells. There is nothing worse than buyers entering your home and smelling something musty. Equally as bad is the smell of strawberry spray freshener or whatever else you bought at the store to make your home air smell better. Buyers are distrustful when they smell air freshener. Remember too that if you are living in the home, you are no longer smelling the air freshener, and buyers may find it overpowering. Yes, this is simple, but it is true. Bake cookies? Great idea. Or heat cookie dough in the microwave. Or turn on a stove burner for 10 seconds and put a drop of vanilla extract on it. A wonderful fresh baked cookies fragrance will fill your house.

Windows. Clean your windows inside and out. Some people advise updating your window treatments, but I wouldn't worry about that. Your buyers will have their own ideas. Clean windows make everything sparkle and they send a subtle and subconscious message to buyers - "well maintained".

Pre-Inspect. As a home inspector, I find that 90% of the time the home seller thinks they know everything that's not right with the home, but 90% of the time they are very surprised when the buyer hires a home inspector to inspect the home and the inspector has an extensive list of repairs. This surprise drives mixed emotions at exactly the time you need to be logical: close to the closing. Often the buyer asks for concessions, money off the deal, and repairs, leaving you, the seller, in a difficult spot. *The money spent on an inspection before your buyers set foot in your home will pay off at closing. Ask the inspection companies you interview for a "Seller's Discount".*

Market. Extraordinary times call for out of the box selling techniques to attract the buyer who will fall in love with your home. Advertising is fine, but everyone else does it. Jump ahead of the crowd with your own Internet web site featuring YOUR HOME. Costly? Not at all. A hosting package with a domain name is $124 a year from the top web firms. Included in this package is a do-it-yourself web site designer. Not handy on the web? Get your child or your neighborhood geek to put it up on the web for a few dollars.

Next, do a You Tube video about your house. Be the star yourself, or ask someone in your family to help you; do one together! Make it fun. The kids will love it. Even a 30 second spot with some creativity will generate visitors to your web site and then to your home.

Follow these tips and sell your home faster, and for more money. Soon you'll be on your way to your next dream home knowing you did everything right. And on top of that you are now an Internet video star.

Take the Home Myth Quiz

Some of the sayings that we hear over and over again may not be true. Can you tell if it's true or if its fiction? Take the home myth quiz and see how you do.

Do not turn off the lights when you leave a room for 30 minutes. Energy surges damage the bulbs. True or False? If you answered False, you are correct. The question was a little sneaky, because the type of bulb makes a difference. Incandescent bulbs (the regular old style we are used to) should be turned off whenever you leave the room, because they waste a lot of energy. The Energy.gov web site tells us to turn off the new CFL lights if you're going to be out of the room for more than 15 minutes. Give yourself 20 points.

Close your crawlspace vents in the winter so your pipes won't freeze. True or False? If you answered False, you are correct. Crawlspace vents are designed to provide humidity and mold control by circulating air. If we close them, the humidity could rise and wood can rot. What about the water pipes? Unless your home is very old, you should see insulated pipes in your crawlspace, or pipes that are close to the warm floor above it and insulation protecting both. If this is not the case in your home, then you may want to close the vents during a hard freeze. Don't forget to reopen them when it warms up. Give yourself 20 points.

Store coffee beans and ground coffee in the freezer to keep it fresh. True or False? If you answered False, you are correct. Coffee beans and ground coffee are delicate and porous. When you remove the coffee from the freezer or refrigerator for brewing, it takes on moisture because of the difference in temperatures. This causes the coffee to lose its flavor faster. The best place to store coffee is in an air-tight container in a cool, dark, place. One exception: if you buy coffee in bulk, wrap it tightly and freeze it. Give yourself 20 points.

If you are replacing your toilet, the best one to buy is the low water volume pressurized type. True or False? If you answered "It depends," you are correct. The two most popular toilet types – split about 50/50 – are pressure assist and gravity feed. In Consumer Reports, both types can be found with high ratings. If you have a big family and hate clogs, choose pressurized; if it's just you and you don't like a noisy flush, choose a gravity feed model. Give yourself 20 points.

Did your central air conditioner spin in? If so, buy the biggest unit you can afford for the best cooling. True or False? If you answered "False", you are correct. A properly sized air conditioning system for your house, based on square footage, will cost less to operate and last longer. In fact, when making a choice between two size ranges – say a 2-ton unit and a 2.5-ton unit, choose the smaller of the two. A larger unit will "short cycle," or turn on and off more often, and will be less likely to dehumidify your home. Give yourself 20 points.

Did you get 100 points? Congratulations!

More Home Myths Revealed

Progress readers asked me for more "myth" questions. If you didn't get 100 points on last week's quiz, try again with these ponderables.

The germiest thing in the whole house is likely to be the doorknob. If you answered false, then you are correct. The germiest thing in your house is not likely to be doorknobs or even anything in the bathroom. If you have a dish sponge that you use for everything from cleaning the cutting board after splitting chicken to wiping the inside of the sink . . . it's full of germs. Then it sits there at the sink rim where bacteria have a party. One of many sterilization techniques is to place it in the microwave (wet, don't wring out) for 2 minutes. Give yourself 20 points.

Radon is a colorless, odorless, tasteless, radioactive gas that may be found in our mountain homes. Radon is continually released by uranium-bearing rocks and soil as the uranium undergoes natural radioactive decay. It is the second leading cause of lung cancer behind smoking tobacco. The best place to measure radon in your home is in the crawlspace. If you answered false, you are correct. The test kit should be placed in the lowest living area of the home. Give yourself 20 points.

True or False? If your home tests high for radon, you will have to move, as there is nothing you can do about it. If you answered false, you are correct. Ventilation systems can be installed to bring radon levels down. Give yourself 20 points.

Remodeling an interior portion of your house is exciting and can add to the value of your home. But don't forget that you will need to go down to the building department and have your changes and upgrades approved before you begin. True or false? If you answered "It depends," you are correct. Painting the kitchen? No permit required. Laying down tile in the foyer? No permit required. Knocking out a wall for an addition with framing, plumbing and electrical? Permit required. Not sure? Give your local building department a call. Give yourself 20 points.

The purpose of caulking is to keep water out of areas where you do not want it. In particular, you should caulk holes in window channels annually. True or false? If you answered false, you are correct. The purpose of caulking is to create a sealed joint between two materials. Caulked joints do help prevent water intrusion, but water should be kept out of the house using metal shields called flashing. Those holes you see in your windows channels are called "weep holes" and are designed to carry water out of the channel, so it's important to leave them open. Give yourself 20 points.

Bonus question. Caulking is applied to the base of toilets in order to keep leaks inside the base. If you answered false, you are correct. Caulking at the base of a toilet is designed to keep liquids from getting under the base from the outside, and making it easier to clean.

Did you get 120 points? Congratulations!

What is Home Downsizing and Should You Consider It?

I was talking to friends the other day about being tired of constantly vacuuming the house. "It just seems like there is too much square footage in the house and it gets dirty too fast," I said. One friend replied, "Exactly, that is why we are downsizing." I asked them to tell me more. "We are going to sell our house and buy a smaller house. We will use the extra money for retirement, and give a lot of things away, and live simply."

This sounded appealing, so I did some research. The Wall Street Journal reports that people decide to move to a smaller home for the following reasons:

- Cost of home ownership. Theoretically, a big house consumes more money in utilities, upkeep, insurance, and taxes. Most people believe that moving to a smaller house will cut these expenses substantially.

- Maintenance expenses and cleaning. Multiple zoned furnaces and air conditioning, landscaping, interior and exterior cleaning expenses can be high with a large home.
- Simpler living with fewer possessions to worry about. At some point in our lives we may move from a "bigger is better" to "simpler is cheaper and more satisfying" mentality. There is something gratifying about consuming less, owning less, and worrying about fewer possessions.
- It is "the right thing to do." As we enter our 50's, 60's, and 70's, we tend to give more away to others who need it more than we do. We are more energy and conservation conscious.

This sounds bright, rosy, and practical. There's only one problem. About 70% of the people who decided to downsize said that it did not work out the way they thought it would. Here's what they did not expect:

- The financial savings was little to none, and some respondents said they actually spent a lot more than they thought they would on the move and in getting the new home ready. They still spent the same amount on television services, telephone, trash pickup, etc.
- Maintenance and cleaning was close to the same amount of money as the home they moved from. Unless you move to a really small home (800 SF or less) these costs only go down a little. Some of the respondents said they downsized by moving from a 2,300-square foot home to a 1,900-square foot home.
- They said that squeezing into a smaller space, and finding new owners for prized possessions gained over decades, was emotionally and physically arduous.
- They missed the home that they left and felt that the new home lacked enough living space and enough storage space. Although the thought of "downsizing" was noble, in reality it was a psychological and financial disappointment. In most of

> these examples, the people who decided to sell their bigger home and move to a smaller home had not done adequate planning or research.

In light of these findings, should you downsize? Yes, if you do it for the right reasons and both understand and prepare for the potential difficulties going in to the process. Next week: how to downsize and not regret it.

Should You Build Your Downsized Home?

If you've made the decision to move to a smaller home, should you build it instead of looking for one to buy? Building your own downsized house might give you exactly what you want without compromise. Consider the following advice if this is your plan.

Advantages of building your own smaller house include having complete control over all aspects of the home design and materials, being able to choose that perfect site to locate it, and integrating the latest technologies in energy efficiency. You can choose your contractor and oversee quality. You can insert some of the things you've always wanted, like a circular staircase, a rooftop patio, a garage or workshop, a deck, or hidden storage.

Disadvantages can be mirror images of the advantages. If you subcontract the work yourself and do not understand the insurance and legal requirements, you can pay fines. If you do not manage the work properly you can end up with shoddy construction and missed project deadlines. If your design is not carefully thought out you can lose energy savings. It takes time to locate the right building site. Space inflation along with change orders as you build can cause the square footage to expand and expenses can skyrocket.

Don't be discouraged. If you are confident that a smaller home of your own dream design can work for you emotionally and fit your lifestyle, then don't be deterred. Think the details through rigorously and then assemble a plan. My advice is draw the home out at the kitchen table, have fun, and then make a list of wants. Prioritize these. For example, "Lots of storage space – 1," Deck space – 3," Loft room – 5," etc. Then find a good home designer. Use the internet and look at their plans. You can save money by purchasing a plan that you like online. Make sure that you have the opportunity to view the plan in 3D and "walk" the home exterior and interior on a computer to make sure it is exactly what you want. Nothing is worse than having a home designed and then finding that the actual layout is not what you thought it would be.

Next, decide where you will not skimp on features. These should include insulation amount and quality, window quality, and appliance efficiency. Use the latest energy efficient bulbs in all lighting fixtures and consider having a blower door test to make sure the home is tight.

Finally, decide who will build it. Unless you are a former builder I recommend against doing your own contracting. You think you will save money by subcontracting the work, but you are well behind the curve that builders have who know what subs are good at any given moment, and have an established working relationship with them. Builders are up to date on the latest materials and methods and can choose the best suppliers. Unless you spend a great deal of time and effort on this, you will end up spending far more time and far more money than you anticipated.

Building your own smaller home can be a wonderful experience full of reward and satisfaction. With the shift in demand to smaller energy efficient homes in the future, the chances are good that the value of your home will go up, not down.

How to Downsize Your Home

Thinking of moving to a smaller home to save money and simplify your life? Think twice. Some 70% of respondents in a survey about downsizing report disappointment and regret. Some of the issues included spending more than the budget allowed, month to month expenses that were higher than anticipated, and difficulty adjusting to a smaller space.

Don't be discouraged. Downsizing to a smaller home can work well, but you need to perform some serious research and soul searching before making the decision. Here is a guide for getting started.

Why do you want to downsize? Reasons can include simply making a change, saving money by moving into a less expensive home, trading a large yard for a tiny yard, moving to a special spot, such as a mountain top or creek, moving closer to family, or simplifying your lifestyle. All of these are great reasons to downsize, but explore and ponder the answers. If the reason is money, do research and make up a realistic budget – you might be surprised. A certified financial planner at this stage can be very helpful.

If your reasons include simplifying your lifestyle, ask yourself if you really can live in a lot less space with fewer storage options and fewer possessions. Do a test – start simplifying now and see how it feels. Put things aside for a spring yard sale; haul things out of closets and make decisions about them. Decide what you have to have and what you can live without.

Downsizing should be downsizing. This does not mean moving from a 2,800-square foot home to a 2,200-square foot home. This is simply a move. Downsizing should mean cutting 30% of your square footage if you really want to have economies of scale. Move from 2,800 to 1,500. Better yet, make a trip to the IKEA store in Atlanta and walk through model spaces emulating homes of 500 square feet and less. A little extreme, perhaps, but it will certainly get the ideas flowing.

Look for or create plenty of hidden storage space in your downsized home. After the move, you will end up with more things than you thought you would. Instead of making your new smaller home look cluttered, hide the extra "stuff" in cabinets, closets, and chests.

Look for or create personal spaces for privacy. A very small home can end up feeling confining and a lack of privacy can be stressful.

Personal space can be very small as long as it is cozy and bright. Look for good examples of this when you visit IKEA.

Downsizing is a lifestyle. I know once in while we think we want to be an acetic and crawl into a cave, having sold all of our possessions. This happens in those fleeting moments when we feel overwhelmed with all the trappings of life and the obligations that go with them. But if you think through what you really want, you can have most of it; peace of mind with fewer things, lowered costs because of smaller spaces, fewer headaches worrying about maintenance and cleaning, and satisfaction knowing that you were able to make that stretch and concentrate on the list of things you really want.

What is a Tiny Home?

Over the past few weeks I have been talking about downsizing your home. This is not appealing to everyone, since most of us get used to living in standard spaces and our inclination is to enjoy larger spaces, not less. The stages of our lives have a lot to do with this. For example, as we grow our families, we naturally want more space and more privacy. When we reach middle maturity, we want the things that we have worked hard for, including the home we have always thought about having. Once we enter our later years, we may want more space for extended family. You may never reach a point where you want to have a smaller home. And if you are in that category of folk who do decide to "downsize," it is even less likely that you will want to downsize to something called a "tiny home."

A Tiny Home is very different from downsizing. In downsizing we move from, say, a 2,800-square foot home to a 1,400-square foot home. If we decide to move into a Tiny Home, we have to either sell everything we own or store everything. A tiny home is either moveable or stationary and ranges from 65 square feet (the size of an average bathroom) to about 500 square feet (a small cabin). Moveable tiny homes sit on a trailer and you drag it to wherever you want it, not unlike owning a motorhome. Permanent tiny homes are sited on property. If you want to have some fun, go to the internet and Google "tiny houses" or "tiny homes." The pictures that will appear are amazing and creative. In particular, the site "Design Boom" - http://www.designboom.com/contemporary/tiny_houses.html - has some very intriguing designs, including houses in trees, houses suspended high over water, and houses built into cliffs.

To actually decide to live in a tiny home, though, is not easy. Even if you can get lined up emotionally with extremely little space, almost no material possessions, and neighbors gawking, you will have trouble installing or building homes this small in most communities. Most homeowner's associations mandate minimum square footage ranging from 1200 square feet to as much as 3,500 square feet and more. Banks will not provide mortgages on tiny homes, and you will need to get permission to set up "house" if it is mobile just like you do with a motor home. Many motor home and trailer parks will not allow tiny homes because they think they look weird. And building departments want nothing to do with these small structures because there is really no way to apply standard codes to them.

So, for now, the only "tiny home" most of us will deal with will be the doghouse or the children's tree house. But look for more news on the trend, which began – where else – in California. In New York City, where square footage is at a premium, builders are getting creative with "tiny apartments" of 200 square feet in vertical designs of 2 or 3 levels. Be glad we live in these mountains.

Are You Ready for an Ultra Tiny Home?

Home building trends from 2008 to 2014 showed a decrease in the square footage of residential family homes. Trulia reports that the average American home size decreased every year since 2008. In the 1950's the average home size was less than 1,000 square feet. By the early 2000's homes were approaching 2,600 square feet.

Whether it was the economy combined with high heating and cooling bills, or the maintenance toll that larger houses were extracting, people began to choose and build smaller homes just before the recession. Good timing. Exceptions abound of course - wealthy buyers are still building mansions, but it is a small percentage of the market. Market share has actually grown for homes sized less than 1500 square feet.

What is most confusing in the data is the "small home" segment. Builders will give you different answers to the question, "How big (or small) is a 'Small Home?'" Consensus is that a "small home" is 1,500 square feet or less. You might say that this is not a small home at all. Then there is the term "Very Small Home." A very small home, if you research the phrase, turns out to be less than 1,000 square feet. Again, you might comment that 1,000 square feet is just fine.

Continuing down this arbitrary path, we come to the term "Tiny House." A tiny house is described by TinyHouseTalk.com as a home with less than 400 square feet. OK, I hear you saying, that is small. But wait. There's more. We have the term "Ultra Tiny House." I bet you can figure this one out. Less than 100 square feet? Yup. These tiny homes can be as small as 90 square feet complete with bedrooms, kitchen, bathroom and a common area. How is this possible? Designboom.com is a web site with pictures of these truly tiny homes and it is remarkable. I don't think I could do it!

Finally, if you want to get away, you can still buy a motorhome or RV with a lot more space than a Tiny House. But you can also get away with a Portable House, sometimes called a "Towable Home." The Custom 30 Foot House by Colorado-based Rocky Mountain Tiny Houses, featured on Gizmag.com, is a handsome example. Built for a family of 3, it offers 200 square feet on the main floor and 100 square feet in two separate lofts. The home features seating for six, a wood stove, large refrigerator, washer/dryer, and a dishwasher. The bathroom includes a composting toilet, sink, a small bathtub, and a shower. The house rests atop a 30-foot trailer, and can be towed by a full-size car or truck.

An interesting final fact on American home sizes. Our homes are considered extravagant by the people in most other countries across the globe. In Spain and France, an average home size is less than half the size of an American home (1200 square feet), and in the United Kingdom homes are one third the size at about 800 square feet. So, the next time you walk into your home, appreciate the fact that it's a lot bigger than you thought it was.

The Lifespan of Your Home's Components 1

Failures in our home's systems and appliances always surprise us. Hearing "Hon, there's no hot water!" or "Why won't the air conditioning work?" feels like getting into our car and seeing the steering wheel missing. This is a tribute to the reliability and longevity of these components. When you reflect on it, you realize that these systems are made up of time limited parts and eventually something is going to break.

When we conduct home inspections, we are very attentive to the age and condition of major systems and appliances. All you need is to move into a new home and have the water heater or AC quit the first week. If you are buying a home, pay attention to these items and use the following information and budget for service, repair, and eventual replacement. Keep in mind that life expectancies of these systems also depend on the level of product quality, the quality of the installation, the thoroughness and frequency of maintenance, weather conditions, and intensity and frequency of use. These guidelines should prepare you better financially and emotionally for the eventual failure of your home systems.

Heating, Ventilation, and Air Conditioning ("HVAC"). Furnaces, both gas and electric, will last 15 to 20 years. Heat pumps last an average of 14 to 17 years. Central air conditioning units last 10 to 15 years. Thermostats are good for about 35 years but you'll want to change them to the latest technology every 15-20 years.

Water heaters. Tankless gas heaters last more than 20 years, while gas and electric water heater tank units last 7-12 years depending upon quality. Electric boilers last about 12 years, while gas boilers typically last over 20 years.

Washers and Dryers. Clothes washers can see rough service and have an average lifespan of only 10 years. I have heard of solid service from some units for more than 20 years however! Dryers last a little longer, averaging 12 years.

Dishwashers last an average of 9 years. Humidifiers and dehumidifiers last about 8 years, and microwave ovens about 9. Refrigerators last an average of 10-14 years, while standalone freezers last 10. Garbage disposals are good for about 12 years.

Ranges will last about 12-15 years, with gas models lasting longer than the electric ones. Exhaust fans last about 10, and trash compacters and standalone icemakers are only averaging 5-7 years.

What this all means is if you built or bought a home about 15 years ago and the systems are all original, get your budget back out of the file cabinet and take a good hard look at what it might cost to replace or repair these appliances, because it's right around the corner. Murphy's Law says that once you do this, everything will continue to work for a long time.

If you get more life from your systems than the above timelines suggest, great! Good maintenance, gentle use, and probably some luck has contributed to this. As an inspector I have seen water heaters test good with over 30 years on them. Eventually though, you'll hear that cry, "Hon, there's no hot water!"

The Lifespan of Your Home's Components 2

In a past column I talked about how long you can expect your home systems, like heating and cooling, and hot water heaters, to last. Let's continue our discussion with non-electronic items. Typically, we think of a home as lasting 100 years or more. This is true – but only certain parts of our homes last this long, as many of us know well when we realize it is time to put on a new roof, install a new floor, or re-build a deck.

Both concrete block and poured foundations should last more than 100 years if they were properly constructed. However, waterproofing materials on foundations only last about 15 years, so inspections are a good idea. Look for settling and cracking – repair or waterproof as necessary. Termite barriers around foundation walls also wear out – check with a qualified pest control company about a guarantee, since termites can do tremendous damage if they gain access to your home.

All natural wood floors have a useful life in excess of 100 years. Slate, granite, and marble are also this durable, but the surface finish will need to be renewed periodically. Vinyl flooring can last over 30 years; linoleum about 25 and carpet will wear out fast: 7 to 12 years is the norm. I am sure you are thinking to yourself that this is a good thing. After 10 years with the same carpet you'll be ready to make a trip to your favorite home improvement store.

Your home's electrical system should last 80 to 100 years if it's constructed of copper. Homes with aluminum wiring have exhibited special problems due to brittleness and corrosion of the aluminum after only 5 to 10 years; if your home has aluminum distribution wiring (aluminum entry wire to the panel is ok) you should have it inspected by an electrician and upgraded. Special electronic lighting controls (Lutron, for example) will wear out after about 15 to 20 years; if you have one of these advanced lighting systems you will probably want to upgrade to the latest technology anyway after that period of time.

Granite, marble, and quartz countertops are at the top of the longevity list at around 30 years, with other materials not far behind (Corian, and laminates) at about 25 years. How well you take care of the materials will make a big difference in the longevity.

Good maintenance and gentle use will extend the life of your home's components and save you money in the long haul. Assemble a checklist of maintenance items, such as air conditioning filters and a walk-around inspection, and run through it periodically to make sure you've got all the bases covered.

How Long Will Your Roof Last?

We expect our home to be one of the most durable things we own. And, like getting into our car and expecting it to start every day, we expect our home to provide a lifetime of comfort and safety. The moment we see water dripping from the ceiling, however, is not a pleasant one, but one that many of us are unfortunately familiar with. Our homes need our attention periodically, and this review of roofing lifespan should help you budget for that unpleasant moment, and hopefully not have to encounter it.

How long will your roof last? Without a good roof over our heads, everything in our home is compromised. First, determine what your roof is made out of. Here in the mountains, the most common materials used are asphalt or fiberglass shingles, and metal. Other types of residential roofing throughout the country include slate, cement, wood shingle, ceramic or cement tile, and copper. Price and durability in our climate usually drive the decision.

Metal roofing typically lasts 20-40 years, depending on the seam type. Asphalt shingle roofing is the least expensive, but the tradeoff is longevity; these roofs will last from 12 to 18 years and the time varies considerably depending on how much sun beats on the roof, and the exposure to weather conditions. It's not unusual to replace one side of the home's roof covering and not the other. Top quality fiberglass composites will last longer, at between 20 to 35 years.

If you are lucky enough to have a slate or copper roof, you may not even need to worry about it for the time you're in the home, as these materials will last 40 to 100 years. Ceramic and cement tile follow at 30 to 50.

Remember that these numbers are dependent on a multitude of factors, including the color of the roof (lighter lasts longer), exposure (north facing roofing will last longer), ventilation (more ventilation and correct vent flow will keep the roof material from overheating), and slope (steeper slopes last longer because they stay cleaner).

Roofers will tell you that "the roof is not weatherproof." What this means is that the roof "covering" is not actually keeping the weather from getting inside your home, the covering is protecting the "substrate" or the material UNDER the roof cover, from damage. The better the roof covering does this job, the longer the roof will last. In fact, if your roof was waterproof, we would experience other problems in the substrate materials, particularly high water content and the resulting formation of mold and rot. Your roof needs to breathe to remain healthy.

Maintenance also plays a part in the longevity of your roof. If you have gutters and they are not kept clear, moisture can accumulate at the roof-line and cause rot and deterioration. In the next column, we'll talk about how to inspect your own roof and determine its remaining lifespan.

Taking Care of Your Roof

The roof over your head is undoubtedly the most important component in your home. Finding out how long the roof will last and knowing the tips and tricks for keeping it sound will benefit your budget and bring you peace of mind.

In the last column, we discussed the different types of roofing materials and how to estimate the life of your roof depending upon what material it is made from. Now let's find out if it is repair time. This process is the same one that home inspectors go through when they inspect a home and issue a report to help buyers decide whether the purchase is sensible or not.

Do you have a pair of binoculars? If so, they will be helpful during this inspection. If you can get on the roof easily and safely, get your ladder out too. If your roof is steep or you don't feel comfortable climbing on to the roof, then we'll inspect it from the ladder and from the ground. If you have windows that give you a view of the roof, this will work well too.

Walk around your home, finding the highest slope from which to view your roof covering. Many mountain homes have a nice upslope that will give you a great view of one side. On the other side, many homes have a deck that you can place a ladder on to view the covering from that side.

Use your binoculars to look at the covering itself – on shingles look for missing pieces, torn pieces, and pieces where the asphalt or fiberglass coating has worn off. Are there any shingles that are curled, blistered, or shiny? If so, anticipate repairs. If you have a metal roof, take a good look at the seams – see the screws holding sections together? Make sure they are not missing and that the washer on each screw is flush against the panel.

Now look at flashings. This can be a leak area. Look at the junction of the roof and the chimney, junctions where a second story begins, and anywhere else where the roof meets another structure. If you see any missing sections or gaps, they should be investigated further by a professional.

Look at the "stacks" as they are commonly called, or the vents, for your plumbing system on the roof. The most common roof leak is around the rubber boots on these stacks. The rubber material deteriorates in the sunlight over a short time – 4-8 years – and will need repair long before your roof covering does. They are easy to inspect – you'll see cracks and voids in the boot material.

Look at your gutters and the wood material at the eave. Are the gutters clean? Grab a section and make sure it's not loose. Take a screwdriver and poke at the wood at the eave. If you see any signs of rot, repairs are in order.

Finally, go into your attic with a flashlight and inspect the underside of your roof. If you see any stains, wood rot, or sunlight coming through where it shouldn't be (light is not uncommon around PVC stack vents but make sure it's not from a broken boot) then you should call a professional to investigate further.

How Long Will Your Deck Last?

Knowing approximately how long major systems will last in your home will allow you to budget for the surprises we have all experienced. If you are buying a home, this information will also help you understand what items need to be renewed, either as part of the sales negotiation process, or after you move in.

The lifespan of your home's decking is probably the most difficult to predict. If you built your deck with engineered (composite, plastic, or metal material) components, then your deck can last 25 to 50 years. Constructing your deck with these materials is expensive but may pay for itself if you can afford it up front, and if you stay in the house a long time. It is more likely, however, that your deck is constructed of pressure treated lumber. In this case, the quality of the workmanship and whether the deck was constructed to code will make a huge difference in its longevity. Weather of course also plays a big part in this equation.

How can you tell if your deck was constructed properly? This is a good question, and a question that you most certainly should obtain the answer to if you are buying a house. Not only does longevity depend upon the quality of the construction, but your family's safety also depends upon it. Some of the decks that I have inspected as part of the pre-purchase due-diligence were about to fall down and there was one in particular that I would not even walk on. I actually gave a warning to both the buyer and the seller about roping off the area until a builder or engineer looked at it to specify repairs. You do not have to be in the process of buying a home to have an inspection of your deck. Call up your local handyperson, building contractor, or a licensed home inspector to come examine your deck. This standalone inspection should not cost you much money and could make a big difference in your family's safety.

Flaws that we see routinely in deck construction include missing hardware, missing supports, missing attachments to the home, missing bracing, undersized beams and supports, and rotted wood sections.

In general, a typical wood deck will need maintenance every 2-4 years (stain or water sealing, nail driving, etc.) and structural repairs every 10 to 15 years. This usually comes as a surprise to homeowners, who are assuming a much longer structural life. But wood, even pressure treated wood, declines rapidly in sunlight and temperature changes through the seasons.

Inspect your deck once a year (or have a professional inspect it and give you a checklist) and keep a high quality opaque sealer or paint on it to make it last as long as possible.

Spring Home Inspection: The Exterior Walk-Around

Along with our glorious and well-spaced seasons come household chores. Are you ready for spring? Most of us love owning our homes, but they do present us with an assortment of problems if we're not attentive enough. And even when we are attentive enough, some bad things will happen anyway.

If you do not already keep a checklist of what to do when, I highly advise it. Type "seasonal chores checklist" in to the internet and you'll find lots of lists to choose from. The following tips are my own favorites; make up your own to suit your situation.

As you proceed with these inspections, write down the things that you need contractors to evaluate. If you do this in in early April, they will not be crazy busy with the summer crush. Many of the items on this list you can do yourself. But if you read the item and say to yourself, "Is she crazy?" then it's time to hire the teenager down the street or a pro to do the inspection (roofs and crawlspaces come to mind).

On a pleasant sunny day take a notepad and walk slowly around the outside of your home. The first circuit should be from about 75 feet out, so you can get the big picture. Take a pair of binoculars with you. On this first pass, look for siding damage, missing paint, gaps, holes, and large cracks; foliage that is growing against the home; damaged downspouts; areas where water is accumulating or running up to the house foundation; view decks for levelness. Now take the binoculars and examine the sections of the roof that you can see. Look for foliage meeting the roof, damaged tiles, and anything that's there that shouldn't be.

Now come up close and make a circuit. Look for the same items as above, and add an examination of soffits, lighting, windows, and deck posts. We'll cover decks in another article, but for this inspection you should make sure that the posts are still sturdy without any decay. Examine all of the vents that you can find to make sure critters and insects can't fit inside any of them. Except for the dryer vent, they should have some kind of mesh or grid to prevent pests from entering. When you come to the dryer vent, make sure the hinged flap closes all the way (spring or magnet). Open the flap and look inside the pipe. Is there any lint buildup? This should be cleaned out – you can attempt it yourself with a good vacuum or hire a pro. Trapped lint can ignite if it gets hot enough.

Finally, take a good look at the grading, or ground slope, next to your home. Does it fall towards your home, or away? In any sections where it appears to lead water towards the foundation, check to see if there's a small ditch five or ten feet from the house that leads water away. If there isn't, then you may be experiencing dampness or high humidity in the home. You may be able to correct this yourself, or you can call a grader.

Keep your checklist going. Once complete, it will make your maintenance life so much easier.

4

Readers' Questions Answered

Readers' Questions Answered - Favorite Gadgets

Inquisitive readers write:

"In your articles over the last few years you have reported on a variety of home gadgets. We would like to know what your favorite gadgets are."

You have noticed that I love gadgets. I spend major portions of my time thinking up new ways to do things, and I am intrigued by electronics and smart home technology. The problem with most gadgets is that instead of saving time and effort, they usually cause you to spend more time and effort fiddling with them when they don't work. Years ago, I installed a whole house lighting control system. 82% of the time the fancy remotes would work. When a thunderstorm came through all of the lights turned on by themselves. This was exciting at 2am.

Over the years, I have learned to do better research when choosing a gadget. Here are some of my favorites.

Have you ever left your garage door open leaving for work? Or worse, leaving for vacation? Or thought you did and turned around to go back home to find that you did close it? It was with enthusiasm that

I read about a "smart" garage door opener – the LiftMaster with MyQ. I have been testing this unit now for six months and it has worked flawlessly. The software is actually an app on your smartphone which means you can be anywhere in the world and be notified if your garage door opens. And best of all – if your plumber arrives ahead of schedule and calls you, you can open the garage door from your phone. All of this high tech does require that you have wireless internet and power on in the home (with a strong password).

I have always loved web enabled cameras. But these can be tricky to get working. One of the first units I tried was at my husband's aircraft restoration shop in 2004. I wanted the people who viewed his web page to see a live picture of the latest biplane project. The setup was difficult and required dozens of calls and experts working over a week to get it working. Since then technology has come a long way on user friendliness and ease of use. My favorite camera now is the Dropcam. All you do is plug it in and download the app to your computer or smartphone. It stores video online, which is terrific if you have a break-in – you have video of it – or need to check if the kids got home from school. I have not had a lick of trouble with it. Once again, you will need a good wireless signal at home.

But my favorite gadget is the simplest. These are the lights that sense motion and come on by themselves. Unlike the wireless lighting mess I got into in the 90's, these are simple switch replacements and have been 100% reliable. Come in the front door at night, the hall light turns on. Entering the utility room carrying things, the light comes on. Starting up the stairs, the light comes on. When the switch hasn't detected motion for a minute or so, it turns off. I am using the Lutron Maestro switches for about $20 each, but local stores have many different models.

Thanks for asking and have fun.

Readers' Questions Answered - The Mountain Stream

Readers call and write to me with great questions. One reader writes: "Since building our home, the basement has always been damp. It appears that rain from the mountain slope is running up against the house and seeping in. Is there anything we can do?"

Indeed, there is. This situation is typical in mountain landscapes where there is no way for the water to get around the house. You should hire a grader to install a drainage ditch on the upslope side of the home where the runoff can be collected and led away from the foundation wall. Putting this about 8-10 feet off the upslope wall of the home will ensure that the water does not seep into the basement. Depending on the size of the area and your landscape arrangement, this may be something you can do yourself. A professional grader, however, will save you hours of backbreaking work (especially if you have a lot of rock in your landscape), and be able to get the drainage just right.

You now have a dry basement, but you also have an ugly drainage ditch next to your house. Is there a way to solve this secondary problem? Yes, there is, and the solution will add value and beauty to your home.

You can do this yourself, or you can hire a landscape company. First, fill your "ditch" with small pebbles and an assortment of various sized rocks. Find some larger boulders to place near the edge. Your "ditch" now appears to be a streambed! Next, enhance this effect by placing mulch along the sides of your streambed, and plant an assortment of plants along your mulched areas. Place a small wood footbridge over your "stream" to your home. Wood bridges are available locally as well as through the internet and are not expensive; you can also find man made rock bridges that look nice. Be creative. Visit landscape companies and the large home improvement stores and look in the garden area for ideas and materials.

If there is still some money in your budget and you want to add some special touches to your streambed and enhance the property further, consider adding a small recirculating waterfall at one end of the streambed. Hook this into a timer so the feature runs when you want it to – say afternoons and early evening. Purchase some 12 volt outdoor lights. These run through a transformer plugged into an outside outlet and include a timer so that they will come on at dusk. These lights draw very little power and will create a garden paradise atmosphere for you and your visitors. You can even purchase a 12-volt light that runs underwater – place this in your waterfall and the effect is complete. Between the sound of gurgling water and the landscape features, your ugly drainage ditch is now a delightful mountain stream. You solved two problems and enhanced the value of your property.

Readers' Questions Answered - Is "Green Building" Expensive?

Readers write: "What is "green" building? Is "green" building more expensive?"

When we say "green building," we typically think of energy efficiency and the use of environmentally friendly materials. However, green building can be expensive in the same way that buying a hybrid car can be expensive – we pay more money up front but will get it back in the form of lower operating costs later in the ownership cycle.

However, "green" building is more than this. What we hope to accomplish in building a green home is more than energy efficiency and the use of recycled materials. Ideally, we want to reduce the amount of resources that the home consumes in all phases of siting, design, construction, maintenance, and end of life deconstruction. At the same time, we want to protect and restore the health of the environment as well the health of the people inside the building. This means avoiding the use of toxic materials so that human health is preserved and the materials don't end up polluting or contaminating the area around the home when it's renovated or re-built.

In practical terms, this means doing quite a bit of research on materials and changing how we design spaces and systems. For example, most carpets contain "volatile organic compounds," or VOCs. These include nasty chemicals like toluene, benzene, formaldehyde, ethyl benzene, styrene, and acetone. Most of us love that "new" smell moving into a new home and don't realize that these compounds are toxic and difficult to recycle. It takes months before these materials air out and the chemical compounds subside. In most cases, carpets will retain some degree of toxicity even after being thrown into a landfill.

Building green also suggests that we spend more time on the building site and the landscaping to minimize impact to existing trees and plants, providing for natural types of water runoff, and installing "xeriscaping" which consist of plants that thrive in the natural environment and do not need extra water and care, thus conserving resources.

So, is building green more expensive? My answer is yes, it is. However, it is not that simple. If you measure the costs during the build cycle, the answer is likely to be yes; if you measure the total cost of the green home over its lifetime, the answer is most certainly no. Green homes will use far less energy over their life spans that other homes, and the overall negative impact on the environment will be much lower.

Should you build a "green" home? If you can find a builder well versed in green building, the answer is most certainly yes. If you try to build a green home yourself you will find it is more work, and probably more expensive than using a professional. The end result, though, will be rewarding in its efficiency, health benefits, and overall low maintenance. If you decide to do this yourself, go to the internet and your library or bookstore and study green building technology until you understand all the components.

Readers' Questions Answered - Front Loaders

A reader writes, "I bought a front-loading clothes washer because I thought it would be more efficient. It does use less water but I have another very unpleasant problem with it. Every time I open the door to put in a new load of wash, the inside smells terrible! My top loader did not do that. What do you think the problem is?"

Here is something that the front load washer manufacturers do not want you to know – front loaders trap water in their seals and over time this water grows moldy and smells. One of the worst offenders is the seal between the front door and the inside of the machine. Many of the manufacturers recommend that you leave the door open so that air can circulate inside, or that you run a load with bleach in it to get surfaces inside sanitized. However, those of us who have septic systems with our mountain homes know that bleach is the enemy of septic system functioning, since septic systems rely on an assortment of bacteria in the processing tank to be effective and healthy. Leaving the washer door open introduces another set of problems - from being in your way and creating a trip hazard to the chance that a small child might decide to investigate the interior. Running a load with hot water does not solve the problem, since even hot water does not always penetrate the seal lip and flush the old water out.

The solution to the problem is simple. Purchase a spray bottle of bleach cleaner, or make up a solution of 50/50 water and bleach in a sprayer. Because this is a strong solution, you should wear latex or medical type gloves, or the waterproof gloves you may use to do dishes or cleaning to protect your hands. Spray this solution on the inside of the washer, covering all surfaces. Take a paper towel and wipe the inside; you do not have to dry it. Keep your face turned towards the outside of the appliance so that you do not inhale the bleach smell, or you can temporarily put on a dust mask. Next, take the paper towel wet with the bleach solution and insert it into the bottom of the rubber door seal. Rotate it all the way around this seal from top to bottom and down the other side. Finally, wipe the inside of the door and any other seals you see. Now close the door and throw your dirty paper towel along with your gloves safely into the trash.

The inside of your washer will be dry and sweet smelling the next time you open the door, and you can leave the door closed until your next load. Performing this operation is quick and easy the next time you begin to smell that moldy aroma, and the minimum amount of bleach that you are using will not affect your septic tank operation. At our house, once I've completed the weekly wash, I perform this bleach wipe in about one minute and close the door, knowing that the following week the inside will be sanitary and ready for a load and I'm not going to bang my knee one more time walking into an open appliance door.

Readers' Questions Answered - Disposals

More good questions from readers include, "Our home builder told us that we could not have a garbage disposal in our new home because of the septic system. Can you tell us why this is a problem? Doesn't the disposal chop the food up enough to be processed through the septic line? Is it illegal? Will the building inspector flag this if we do put one in?"

Builders and plumbers are telling you not to install a food disposal (also known as an "insinkerator") because they know that many of us may be putting large amounts of food and plant matter down the drain. Some of us have moved from a geographic location where homes are on a city sewer system and folks do not think twice about putting the strangest of items down the drain, including coffee grounds, cooking oil, and cigarette butts. Refusing to install a disposal sends the message that things are different in the mountains. And indeed, things are different here. Some of our septic systems – particularly the "drip" system, which has a special drain field – require great care on our part to prevent clogging and maintain healthy processing over the long haul.

It's not illegal to install a disposal under your kitchen sink, and the building inspector will most likely not be concerned about it. However, if you do decide to install a disposal, there are practices that you should follow so that the "stuff" you put into the disposal does not create an expensive septic system headache for you.

Use your disposal ONLY for the debris on your plates and dishes that is left AFTER you scrape the bulk of the leftovers into the garbage container. You may think that this defeats the whole purpose of having the disposal, but I can tell you from experience that this is a nice feature even if you can't dump everything down the drain – if you put small amounts of food into the disposal, everything will be fine. It is also convenient for soups and liquids that are difficult to place into the trash. Never put lots of plant or vegetable matter into the disposal – this includes potato peels, celery, broccoli, etc., as the septic processor has a hard time breaking down this material in the tank. You should also avoid pouring any leftover oil from cooking down the drain; pour this off into a container, or wipe small amounts out of the pan with a paper towel – and place in your garbage bin. We like to keep a small plastic bag in the freezer that we place bulk food matter in to, and then toss it into the garbage right before we take the trash to the curb or the trash station so it does not decay and smell in the interim.

Finally, what do you do about that certain smell that a disposal develops over time even when you rinse it with detergent and hot water? Simple. Keep an old dish brush handy under the sink and give it a good scrub inside – inside walls and metal blades - with dish detergent from time to time. Then run it with a handful of ice cubes and a small lemon wedge with hot water to make it fresh smelling.

Crawlspace vents – should they be closed in winter?

A Progress reader writes, "We have sealed up the drafty areas in the house and it feels a lot cozier. Our floors are cold though. We have a dirt crawlspace and we don't know whether to close the vents or not in winter. What do you advise? We don't have any insulation under the floor, and the space is just bare dirt."

I get this question all the time on inspections and it's a good one. There are several things you can do, depending on how much or how little money you want to spend. At a minimum, I would cover the dirt in the crawlspace with 12 mil plastic. This will stop dampness from rising out of the ground and penetrating your floor structure, and it will keep the space warmer. You can get this "vapor barrier" at local home and hardware stores or online at www.crawlspace.net. If your crawlspace is fairly dry and humidity is not creating mildew on your floor under-structure, then I'd close the crawlspace vents to help retain heat. If you can spend the money, I would then add insulation to your under-floor area. This one step will make a large difference in how well your home retains heat (and reduce energy costs). If you want to go one step further, close off your crawlspace vents completely and add a dehumidifier to the space. Technically, crawlspaces and attics are supposed to have a certain amount of ventilation, and building inspectors inspect to these standards on new homes. The "green" building movement has produced some variations, however, that contribute to energy savings but have to meet the inspection requirements for ventilation – hence the addition of standalone recirculating systems. These systems keep the structure clean and dry and mold free. On some inspections, I have crawled through spaces that were so damp that water was dripping off floor joists, floor boards were rotten all the way through, mold was growing on the wood structure and the walls, and I was crawling through puddles (in a waterproof suit!) throughout the space. The owners complained about cold and damp floors and did not realize there was a problem until a member of the family stepped on a rotten section in a closet and ended up in the crawlspace! If you have these kinds of problems, I would definitely hire a qualified contractor to review your situation and advise on next steps.

If your crawlspace is humid and damp now, I recommend solving that problem first, before you close the vents. Adding the plastic to the dirt may be the complete solution, but try it and see. Leaving the vents open in winter will not hurt anything and will probably help lower the humidity by introducing ventilation. The downside is a colder crawl-space, and potentially frozen pipes if your pipes are not currently insulated in that area. Ideally, you can add insulation under the floor AND the vapor barrier, and close your vents in the winter if you find that the humidity is 60% or less. This will keep your floors warm, and your pipes safe.

5

All About Decks

Spring Safety Checklist for Your Decks Part 1

When you first arrived in this glorious mountain paradise – no matter what direction you drove in from - you probably noticed the expansive decks on our homes first. Blessed with a moderate climate (although this past winter might cause you to disagree with this statement) and four wonderful and distinct seasons, our decks can be used nearly year-round to enjoy the forest, lakes, mountains, and wildlife along with the spectacular views that they provide.

Decks invite us to have family and friends share in the experience. Decks are our respite from the stresses of daily life, bringing us time and space in which to re-center our thoughts and emotions and take a deep breath, appreciating all that we have.

If you have a deck attached to your home (and 90% of us do), you may have noticed that they need constant attention. Most of our decks are made of pressure treated wood – a material that looks fabulous for about four months after being built, and then suddenly time warping to an age of about 5 years – or so it seems - as we see the cracks and twisting, and the stain turning lighter and thinner. Knotholes open up, nails loosen, boards warp, and we wonder what happened. If you're identifying with this experience, you are not alone.

While our homes can be sturdy for a hundred years or more, the decks attached to our homes have a lifespan of about 12 – 18 years before they need structural maintenance. Putting a good opaque paint on the wood will prolong the life of the deck, but some parts and components can't survive the beating from constantly changing weather conditions.

In addition to weather causing general deterioration, a majority of decks are not built for longevity. Almost like an afterthought, the decks go on last, and important details may be left out. Two examples of important detail that are routinely left out are permanent footers for the deck posts and not attaching the post to the footer. After 6 years of ground settlement, the owner wonders why the outside portion of the deck has dropped, only to discover that the footers have cracked and the posts have moved. They are then shocked to learn that they have some major repairs to a relatively new home.

The Consumer Product Safety Commission (CPSC) conducted a study of deck-caused injuries to people from 2003 to 2007. An average of 44,800 people a year suffered an injury related to a deck or porch according to the study, and of these 15% were directly related to the structural failure or collapse of the deck or porch. Data suggest that over 75% of these structural failures happened at the deck-house connection, or at the part of the deck called the "ledger board" which is the band of wood on the inside portion of the deck that attaches to your home. It has been common practice to use nails to fasten this important structural component to the house. We know that nails can pull out, and with the safety data collected over the years, most municipalities have adopted some version of the International Residential Building Code which requires that deck ledgers be mechanically attached (bolts, not nails) to the home. Your home inspector may not flag your 1996 nailed deck as a repair item because at the time it was built, bolts may not have been required; but he or she will certainly point out the safety hazard and the need for regular inspections. The good news is that most decks don't last longer than 15 years without something needing to be upgraded or repaired, so this is the opportunity to have the structure brought up to date.

Next week we will begin a series on deck safety and we'll talk about what you can look for to keep you and your family and friends safe on your decks. You will get to play "inspector" and we'll use a handy checklist to inspect your deck that you can use every year to make sure all is well. If all is not well, then you'll know what's wrong and how to fix it.

Spring Safety Checklist for Your Decks Part 2

Spring is springing and it's time to spruce up our decks and porches! After a tough winter in the mountains, we're ready to go outside and enjoy the many reasons we came to this area –forest breezes, clear fresh air across the mountain and lake vista, and a variety of wildlife visitors to our back yards. Where do we go? Our decks, of course!

Our decks – whether small cantilevered balconies overlooking the lake or expansive multilayered outdoor rooms with roofs over them – are an important part of our homes. We think they should hold up over the years the way our house does. But our decks have two characteristics that create problems. They are often not built correctly (or safely) and they are exposed to some of the roughest weather imaginable. This past winter is a good example. Freeze, thaw; freeze, thaw; wind, rain, sun, more freezing cycles – you get the idea.

This article begins a series on deck safety. Along with trips and falls in the home, decks can pose a multitude of opportunities for injury even when they are built correctly. When they are not built correctly and have not received maintenance, then the injury possibilities multiply. Let's take a tour of your deck.

Inspectors first walk the perimeter of the home on the outside from about 50-75 feet away. This gives us a good perspective to find structural issues and get acquainted with the layout. Then we circle the house at 30 feet. We circle once more up close from ground level. Then we go underneath the deck and examine the components where they meet the house and the structural support from the inside to the outside of the deck. Finally, we go on to the deck and look at house connections, railings, stairs, etc. Let's go! Come with me. We'll start at the 50-foot view outside.

Do your posts have straight line load support? Take a look at the vertical posts that go into the ground. If you only have one lower set of posts that go from the ground to the deck floor and nothing more, then you can skip this. If you have a roof over your deck, or more layers of decking, then this will apply to you. We want to see that the vertical posts on the deck above line up with the tops of the posts that go into the ground. The reason we are concerned about this is because structure is designed (by engineers) so that loads sit on parts that are capable of holding them. When the upper posts are not over the lower posts, then other parts of the deck are trying to carry a bigger load than what they were designed to bear. This means that parts can break, twist, or fail. Multiple posts that are substantially out of line (more than a 45% angle away) with upper or lower posts or columns should be reviewed by a licensed contractor or engineer depending upon the complexity of the deck.

Don't call the contractor yet – we'll continue our tour next week!

Spring Safety Checklist for Your Decks Part 3

Last week we talked about deck safety and how you – the homeowner – can inspect your deck yourself – "play inspector" – and find out what to do if you find any problems. The good news about decks is that it is easy to determine if you have a problem, and problems are relatively easy to fix.

Come along with me once again. We're outside now on our up-close walkaround. Let's look at posts and footers.

Footers. What in the world are footers? They are the anchor or structure at the bottom of your deck posts that keep the structure from moving around. Are the posts that support your deck secured to concrete or masonry blocks that go into the ground below the frost line (12-16 inches)? We may need to dig down along the post to see if they do. The posts should have footers, especially if they support house or roof structure above the deck. If you discover that there are no footers, or the footers you see do not go into the ground (like a 4-inch pad) then you may have a problem with the posts settling or moving as the material reacts to freezing and thawing throughout the year. If you do have temporary footers, then you should monitor them over time for deterioration or settling. If you notice any movement, you should have a contractor reinforce or replace them with permanent footers.

Are posts positively attached with a bracket to the footer? You and the inspector may or may not be able to determine this, if the post goes into grade (soil). When the post sits on a pad, we look for a bracket that secures the post to the footer with fasteners, which can be a variety of bolts, screws, or the manufacturer's recommended fasteners. Posts that are not attached to footers can move over time, damaging other parts of the deck structure. A contractor can make the necessary connections.

Are all posts relatively straight (not excessively warped) and are they free from obvious movement in, out, or sideways? If not, they may have moved due to settlement since it was constructed, or they were already warped when they were installed. This may or may not be a defect; if the deck has settled slightly and is now stable, it may not move again unless there is additional instability in the soil. Pressure treated lumber does warp, but excessive warping will pull fasteners out and the post will twist and move, exposing the top grain to moisture damage and excessive wear. Excessively warped or leaning posts should be repaired or replaced.

In ground posts. Most decks are constructed this way. Even though pressure treated wood is designed to withstand weather and soil contact, you should inspect the sides of the posts where they enter the soil for any signs of rot. Use an awl and probe the wood below the soil. Dig down a little and take a look at the condition of the post. If you discover softness or dry rot where the wood pulls away easily, or holes in the wood, you should have a contractor replace or repair the damaged sections. Also, try to keep your downspouts away from the soil around your deck posts. The soil has enough moisture without adding to it with water pouring on the posts – this will rot them out faster.

Next week we'll inspect the upper sections of your deck. Stay tuned.

Spring Safety Checklist for Your Decks Part 4

Last week we continued our discussion of deck safety and how you can inspect your own deck for potential problems. Let's continue with our walkaround and look at deck support and structure.

We are still in your back yard looking up at your deck. Do you see the rim board? This is the horizontal band of wood that circles your deck between posts. It unusually consists of "two –bys" which is contractor-speak for a two by eight, or a two by ten, etc. – boards that are placed horizontally on their sides and often doubled or tripled up to make a structural beam. Make sure there is no sagging in this board or boards and that any seams you see are not pulling apart. If you see any joints opening up or any sagging, mark this down for maintenance.

Lateral support. Are there any support "arms" that extend from the deck post sides at an angle back up to the rim board? Some people call these supports "crow's feet". These provide strength to your deck and keep it from twisting and moving. You may or may not see these on your deck, depending upon the height of the posts. If you do have these supports, then they should be bolted, not nailed, to the structure (rim board).

Anti-racking. If you have lateral supports on the posts, you may not need bracing on the under-deck structure. And similarly, if you have under-deck bracing, or what is called anti-racking (anti-twisting) support installed under-deck, then you might not need the lateral support arms. These items would be specified in the plans for the deck and are dependent upon the required engineering elements for the strength of the deck. If your inspector does not see either one of these elements – anti-racking or lateral support arms – he or she may suggest that you have a contractor or engineer review the structural components for adequacy.

Now let's take a look at the structure up close. We will need to walk under the deck at all levels. We'll start at the ground level.

Is the deck bolted to the house? Take a look at the board that attaches the deck to your house. Are there bolts attaching this board, called a "Ledger Board" to your house? If it's only nailed in, these nails can pull out, taking the entire deck with it. Of all structural failures in decks, this is the most common, and the most serious. The good news here is that if your deck does not have bolts, it is a straightforward job for a contractor to install them.

Hangers or ledger strips. Now, since we're already under your deck, take a look at the support boards, called "joists" that hold up your deck planking (the boards you walk on). How are they attached to the ledger? You should see joist to ledger "hangers" or metal brackets that hold these joists to the ledger, and you should see them where the joists connect to other support structure. These brackets will have nails and every hole on the bracket meant for a nail should have a nail in it. Some decks use a 2X2 inch wood strip under the joists to help secure them (instead of a hanger) and this strip is nailed to the lower portion of the ledger board and to the outside band (rim) board. Ledger strips should be firmly attached – nails ok – make sure there are sufficient nails (3 in the vicinity of each joist is good). Take a close look at these areas and make sure nothing is pulling out or pulling away. Hardware you see that looks rust damaged should be replaced.

Spring Safety Checklist for Your Decks Part 5

Let's continue our discussion of deck safety and how you can inspect your own deck. Did you know that May is Deck Safety Month? It's a perfect time to follow me around once again with this checklist. At the end of this series I will tell you how to get a copy of the entire checklist in case you've missed any articles.

Believe it or not, we're still walking around under your decks looking up at structure. Let's look at a few more things and then we'll go up on to your deck and look around.

Last week we talked about how you should be looking for "hangers" or brackets on the ends of the joists, OR a ledger strip – a 2X2 piece of wood under the joists. If you have hangers, take a good look at them. Does the bracket stretch all the way from the bottom of the joist nearly to the top? What? They only go half way up? Hangers should be the right size for the joist it is holding up. If the hanger is short, you will not get the required strength that was specified in the plans. A similar situation exists with missing hangers; they might be the right size, but not all of them were installed. In the case of missing hangers, you should install them where they are missing. In the case of short hangers, you should inspect your deck a couple of times a year to make sure nothing is pulling out or away.

Joists bearing on structure. Are all joists where they sit on top of posts attached positively and are they on the post with at least one and a half inches of contact? If not, the joists are not as secure as they should be, and they could pull away. You should add reinforcement as necessary.

Missing or improperly installed flashing. Galvanized metal, plastic, or similar material that will resist corrosion should be used between the deck and the house, so that water does not run into the wall, or contact any non-treated lumber. You will see the flashing where the deck meets your house as a material that comes about 4 inches out from between the deck planking and the top of the ledger board. You may or may not be able to see this from the top side of your deck, as the flashing should be under the siding and planking but over the top of the ledger to lead water away from the wall. You should be able to see it from underneath your deck. If flashing is incorrect or inadequate, moisture from rainfall can get into the wall and siding material and cause deterioration and eventually wood rot. If you do not see it, or sections are missing, you should have a contractor examine the areas and make a recommendation on how best to fix it.

Finally – let's walk around to your stairs and go up on to your deck. If you have multiple levels of decks, you will want to repeat this process with the other levels.

Grab the post that the stair rail is attached to and try to wiggle it. It should not move. Do the same with the railing. It should be strong against your push. Railings should be able to hold a lateral force of 200 pounds – if you find your stair rail wobbling, it should be reinforced. Now look at the stairs themselves. Are they secure? No wobble? No nails sticking out? No tripping hazards?

Next week we'll finish up our walkaround and your deck will be ready for play, relaxation, and entertaining – safely.

Spring Safety Checklist for Your Decks Part 6

In the home stretch! This is the final article in the deck safety series – we'll complete our inspection and you'll be a deck safety expert! This month is "Deck Safety Month" so the time is right to get everything shipshape and ready for summertime. Last week we started up on to your deck via the stairs (if you have them) and you checked them out for looseness, and falling and tripping hazards. Now we're on your deck and looking at the railings and posts.

As you did with the stair rails, hold of the top railing and try to move it. It should be strong and secure, and not move against your weight. If it moves, it could mean that connections are loosening up and should be tightened or reinforced. Remember we said that railings should be able to withstand a force at the top of 200 pounds in any direction. Also, all decks that are over 30 inches above the ground should have guardrails.

Now look between the horizontal rails at the small posts filling in that space – we call this "infill". Is the spacing of the pickets, or vertical posts that keep children from falling through this area, 4 inches or less? You should also look at the space between the bottom rail and the deck. This space should be no more than 6 inches. This standard is relatively recent, so your home may not have this spacing. That's ok. But when you replace your deck components, or make improvements, you should move to the new spacing for additional safety.

Now look at the surface condition of your deck. Is it worn? Are there nails sticking up? Splinters? Take a small screwdriver or awl and poke at the areas that look damaged or soft. If the blade of the screwdriver easily penetrates the wood, or the wood easily flakes off, then you have rot damage. These sections should be replaced because they have lost their strength. Look at the planking you walk on. Are there any big knotholes that could catch a shoe heel (ok, I know we don't wear high heels in the mountains, thank goodness, but you never know when Lady Gaga or Jennifer Anniston will visit you).

If you have pressure treated wood on your deck rather than a composite (such as Trex) you'll want to seal it regularly so it lasts longer and looks good. As you may already know, this is no small task. If you do not mind a solid color, painting your deck will give you longer intervals between treatments.

You're done! Congratulations! But wait. Now that you have your "list", what are you going to do? If you've got multiple "repair" items on your list, you are in the majority, not the minority, of homeowners.

The good news is that most deck repairs, with the possible exception of inadequate or failing footers, are straightforward and not overly expensive unless you are replacing major sections of structure. More often, a good contractor can repair your deck with a minimum of time and money. Many of the items we've explored in this series are easy to do yourself – like adding or replacing hangers.

Now that you're the expert, inspect your deck at least every spring, and keep a good sealer or a good paint on the surfaces. This will stretch out the time between component replacements and ensure ongoing safety and enjoyment.

Spruce Up the Deck

Winter here in the mountains was as tough on the exterior of our homes as it was on us. You may have noticed that your decks aren't looking as good as they did last year. If you're lucky enough to have a good synthetic deck with color built in to the boards, you are fortunate. Most of us have regular pressure treated pine on our decks. PT decks, as we call them, are reasonably priced and very durable if cared for. Here are some tips to bring your PT deck back up to great.

- Safety first. The first step is not counting how many nails have pulled out on the deck surface or how many knots have fallen out. The first task is going underneath your deck to make sure all the hardware is still attached. Look at supports and beams, and grab posts and apply pressure to see if anything is loose. Now go up on to the deck and push against railings and posts. If everything is tight and strong, we can proceed to the next step. If your deck needs structural attention, go ahead and get those items done first.
- Cleaning. Yes, I know, this is not the fun part. But preparation makes or breaks a quality job, and a quality job can greatly extend the life of the finish. Take the necessary time to do

this right. If it sounds too arduous, find a good handyperson. If the deck is relatively clean, you can buy OxiClean or other outdoor cleaner, and use a stiff bristle broom to apply it. If the deck has mildew and lots of dirt, a pressure washer is effective with a deck detergent or a diluted water and bleach mix (80% water, 20% bleach). For this cleaning stage read the directions carefully on the cleaner, and wear eye protection, gloves, and old clothes that protect skin.

- Fixes. Once the deck has dried, inspect it for popped up nails, knotholes, and cracks. If you've got a hole in the deck you can nail a small piece of wood underneath it and then fill it with structural epoxy (glue mixed with a filler), or even wood filler. If you have a board that is warped and nails won't keep it down, try some long drywall screws. In fact, your next deck should be screwed down, not nailed. No pops and no warping. Yes, more expensive – you decide whether it is worth it or not.
- Choosing a stain. No problem picking colors – this part is fun – but how do you know what to buy? Two factors – transparency and brand – drive longevity of the finish. Solid stains will last longer, but you won't see the wood grain, which most people like to see. Brand can make a huge difference, so choose the top picks from Consumer Reports magazine. They test paints and stains over many years and evaluate resistance to mildew, dirt, cracking, color changes, and overall value.
- Apply the stain. It is critical that you allow the deck to completely dry out before beginning your application. Don't believe what you read about being able to put product over damp wood. You'll be sorry! Although spraying stain on is easier, brushing it on will give you a longer lasting finish.

Have fun!

How to Waterproof Under Your Deck

If you have an open deck with another deck below it and/or a deck with a patio below it, you are used to dampness, mildew, and the inability to go out and sit on that lower deck or patio when it's raining. Sometimes the best time to enjoy our mountain environment is in a gentle rain – but not when we are getting soaked. If the storage area for your car, tractor, or garden equipment is directly under a wood deck, you are familiar with the mold and wet conditions that make it difficult to keep things dry and clean.

Over the years as a home inspector I have seen the lengths that people will go to keeping rain out of these areas, sometimes creating a problem that is worse than the one they started with. One homeowner stapled plastic to the underside of the deck to protect a carport area. The plastic was a mess. Pockets of brown water bulged down, with dozens of leaks. From above, the smell of mold and decay rose up from between the deck boards. Needless to say, I advised the buyer to rip out this ill-conceived solution.

What should you do if you like the idea of running water away from an under-deck area? It actually is possible to do this and not have it leak or produce mold. Keep in mind, though, that anything attached to your deck will take on the high maintenance characteristic of a deck. This means you'll need to clean leaves and debris from any water runoff system by cleaning and flushing it from time to time.

There are four systems that will keep your under-deck area relatively dry: waterproof membrane, between the joists metal or plastic, below the joists attachments, and do-it-yourself. The first and second methods are employed when you're building the deck. They are typically installed before the deck boards go down. Since we are talking about decks that are already installed, I'll skip these first 2 systems. At the end of this article there is a link if you want to read about all of the methods.

The third system involves installing fitted plastic sheets underneath your deck on a slope with a gutter. This one is my favorite because it's simple. One of the best off the shelf systems is called DrySnap. It's available at local retailers and is straightforward to assemble. One of the advantages of these "sloped roofs" underneath you deck is there is less entrapment of the junk that falls between your deck boards.

The fourth system is a self-designed and made rain catcher. If you're a do-it-yourselfer and are familiar with deck construction, have fun and save money. I have seen some good ones on inspections, but overall 70% were trapping mold and debris and had multiple leaks. So, if you do design your own, think the project all the way through and do your research.

To look at the possibilities, go to www.deckmagazine.com and type in the search box: "under-deck draining systems," and you will see a pdf document with this title. It provides a complete evaluation of all the systems to get you started.

Tips for Deck Care

There is one item attached to our house that can always use some maintenance: the deck. If you're fortunate enough to have a deck, read on. If you're fortunate to not have a deck, then read on so you realize what the rest of us have gotten ourselves into.

Most of us who have wood decks would probably say that the pleasure of a deck space far outweighs the inevitable repairs and maintenance. This doesn't mean that we don't grumble and resist when these chores move to the top of the list. If it makes you feel any better, you should know that putting a deck on your home will return about 75% of what you spent – making it one of the top cash returns at resale.

Consider the following 5 wood deck care tips so you can spend more time relaxing on your deck knowing you have it under control.

- Use a checklist. You can get a maintenance checklist online or at your favorite builder supply, or you can make you own.

Divide up the chores throughout the spring, summer, and fall so you don't feel overwhelmed. If you try to tackle all of them on one long weekend, you will be frustrated.

- Use the most durable surface finish. As much as we love the see-through wood grain of transparent and semi-transparent stains and water-proofers, you can face a lot less work year after year if you go with a water based solid paint. Stains do not protect the wood as well as paint from UV rays and moisture, which are the two universal outside damage deliverers. The opaque finish of a solid outdoor paint will hide imperfections, reduce mold and decay on the wood surface, is highly cleanable, and will last 5 times longer than a transparent stain. Less work or better look? You decide.
- Address mold and decay. Our rain soaked spring and summer have delivered the worst possible conditions for decks. If sections of your deck are in shade, it's a double whammy with dirt, dust, moss, and molds finding their perfect growing environment. When it's not raining, make sure your deck is swept clear of leaves and debris. To clean off mold, use vinegar and water mixed 2:1 with a pressure washer. If the mold is extensive, you may want to have the deck professionally cleaned.
- Identify structural problems early. Walk into the back yard and look at your deck from about 30 feet away. How does it look? Everything straight? Our eyes are good quality control devices. You can tell a lot from this perspective. If you see anything leaning or broken, you should call a professional.
- Do a safety check. Grab all railings and shake. They should feel solid. Do the same with stair railings. Go under the deck and check for loose or rusted hardware.

Our decks and outdoor spaces are meant to sooth away the cares of the day and bring us some perspective as we observe the slower pace of nature. These tips should help you spend more time relaxing and less time maintaining.

To Pressure Clean or Not To Pressure Clean?

Are you running out of time on your summer projects? Do you have a list of things to accomplish before the weather turns cold? I am, and I do. The days are getting shorter, and it definitely feels that way when you're trying to get things finished on a deadline.

This year it is decks. Procrastination is easy when you consider the work involved. We all know that preparation is critical to how the job turns out; a poor prep job can negate days of finish work. But it's also human nature to try to make the preparation step shorter or simpler.

Enter pressure washers. Should you use one to clean your pressure treated pine wood decks? One reader told me that directions for re-staining her deck said to use a pressure washer to make the cleaning job easier, but then when she went to the store to buy cleaner and stain, the pro there told her to not use a pressure washer.

To use a pressure washer or not is going to depend primarily on 3 factors. The first is the type of washer you have or are renting. Residential type washers range from 1,000 to 3,000 psi (pressure per square inch, or force). The second factor is the type of nozzle you are using on the washer – nozzles come in a variety of flavors, from a wide fan pattern to a round single blasting stream. The third factor is your own level of experience and training using these washers. An inexperienced person with a 3,000-psi washer and a blast nozzle can do a lot of damage in minutes. This setup is so strong that it can rip out chunks of wood, and if pointed at the house, break windows and crack siding.

The reason that store personnel tell you to not use a pressure washer to clean your decks is because they know that whatever advice they give will come back to bite them if anything goes wrong with your job. They are treading a fine line between helping you get your job done right and keeping you out of trouble.

So? Should you, or shouldn't you? My advice is to go ahead and use a light duty electric or gasoline pressure washer with a low pressure setting and a fan spray pattern nozzle. If you have little experience with one, then get someone to show you how it works. Remember that pressure washers are not designed to remove paint or mildew; use a commercial cleaner or stripper with a bristle brush and protective equipment for these, followed by pressure washing. Follow the directions explicitly.

If you don't mind applying a little elbow grease, forget the pressure washer and use your garden hose with a good spray nozzle. Get a long-handled bristle brush and a five-gallon bucket. Pour 2 cups of Oxi-Clean or ordinary bleach in your bucket with 3 quarts of water and a ¼ cup of powdered laundry detergent. Using the brush, apply the mixture to the dirt and mildew and let sit a few minutes. Scrub in several directions. Rinse thoroughly with your garden hose and allow to dry thoroughly before staining or painting.

Deck Safety in the News

Over the Fourth of July weekend many of the local news stations ran stories about deck safety. It's not something we usually think about, but it should be at the top of your maintenance checklist, especially with summer in full swing. This is particularly true if you're inviting a group of guests and family members over and you'll all be out on the deck at the same time. Many of the deck failures involve group photos. What's bad about group photos? Well, this is the chance to load your deck to its maximum and test whether the structure underneath your deck will actually support the weight of all these people concentrated in one spot on the deck. What could happen? The flooring gives way under the group, or the deck can detach from the house – both bad news.

Hundreds of injuries are caused every year by decks that have not been built correctly, or are old and uninspected. Home inspectors will tell you, and it's been my own experience, that 50% of all decks have been built incorrectly and/or are structurally deficient. The number of injuries nationally is going up, not down.

Here's how to check your deck to see if the group photo is safe. If you have any question about your deck's safety, call a licensed home inspector or contractor to inspect it.

- Wrong or missing hardware. Look for nails in every bracket hole, and a bracket (called a "bucket") under every joist, or wood member, attaching it to the main beam under your deck.
- Temporary piers under deck posts. Do you see a concrete block (called a "pre-cast foundation block") sitting on top of the soil with the post sitting in it? A result of lazy building, these blocks under your deck posts can, and do, move. I've seen some posts out of plumb by as much as 20 degrees. Here in North Carolina, posts should be attached to a concrete pier that goes into the soil below the frost line. In Clay County, this is 4 inches. Securing the pier below the frost line prevents movement, and setting the wood post on a pier above ground helps prevent rot.
- Incorrectly attached deck ledgers. The deck ledger is the board that everything else on the deck is fastened to, securing the whole assembly to the side of your house. To inspect, go under your deck and look for large bolts (not nails) at regular intervals. Ledgers should not be fastened to brick, stucco, or foam without specialty hardware, because they can pull out. Statistics indicate that 90% of deck collapses in the last 10 years occurred as a result of the separation of the deck ledger

board from the side of the house, allowing the deck to swing away from the house.

- Supporting posts bolted to the sides of beams. Posts should be under structure, not alongside it. The bolt will not fail – the wood will.
- Incorrectly attaching deck guardrail posts. A guardrail by code is supposed to be able to withstand a load of 200 pounds in any direction along its top. Would you be afraid to try this test on your own deck rails? If you find your rails the least bit wobbly, get someone to install extra hardware correctly to make it safe.

Conduct this deck check yourself, or get a professional to check it. Then go for the group photo!

6

Home Automation, Gadgets, and Security

Cooking Gadgets

I am not a cook, and I attempt to spend the least possible amount of time in the kitchen. I have great respect and admiration for those of you who love turning out wonderful home cooked meals. Since I didn't get the cooking gene, I look for ways to save time producing a meal. And since I'm an engineer, I try to maximize the number of gadgets in doing so.

One gadget that I love is the microwave oven. I actually need several microwaves, and if my family and friends didn't tease me so much I'd have a third microwave. So, when I saw an ad this last weekend for something called a "NuWave Oven" (Amazon.com, $139) a device that claims to dramatically shorten cooking time and produce delicious food, I figured they designed this gadget for me.

The NuWave oven claims to combine 3 distinctly different methods of heating food – conduction, convection, and infrared. Conduction transfers heat directly to food; an example of this is when you place a burger into a frying pan. Convection transfers heat to food by circulating heated air around food. Infrared transfers heat to food by surrounding the food with very intense heat waves. Where microwaves cook food from the inside out, infrared energy cooks from the outside in, much like the radiant heat we are used to in our conventional oven. Infrared waves are sandwiched between the visible spectrum of light and microwaves. The reason we are hearing about infrared now is due to a patent expiration in the year 2000. Since then scientists have been improving the technology and bringing the cost down.

While most all of us recognize and are comfortable with conduction, convection, and microwave cooking, infrared, and even another technology – induction – seem a little baffling at first. Why would we want to even consider new tech? We certainly can stick with what we know, but the gadget geek that I am makes me want to find out more about these new technologies. Here's what you need to know.

Both induction and infrared heating methods are very, very fast, and get very hot. Induction cooktops will boil water in seconds with a surface that is cool to the touch. Rather than using electromagnetic waves through the food, induction technology uses actual electromagnets under the burner that transfer heat to the pot or pan. Drawbacks: these cooktops are still quite expensive and you can only use ferrous (metal) pots – no aluminum or ceramic materials will work.

The price of infrared cooktops is still high, but infrared grills have come way down in price to the point where they are all the rage in some areas of the country. Because the heat is from the outside in, infrared provides a nice searing of steaks for example, and locks in flavor and juices. Drawbacks include the possibility of overcooking food and the scarcity of repair people if you have a problem.

Now that you know all the answers to the latest cooking technologies, should you go out and buy the NuWave Oven, which combines in one single countertop device three of the above technologies? After I buy this gadget I'll let you know. In the meantime, I'll research that third microwave.

New Year Organizing Tips for the Home

Because it is the time of year when we are making outlandish New Year's Resolutions, I thought I would talk about realistic organizing techniques around the house. I am seeing articles this time of year that go into great detail discussing how to make your junk drawer look neat. While I am definitely a person who dislikes clutter, I think that de-cluttering a junk drawer is over the top. Here are some realistic ideas for tidying up your home.

- The entry to your home. Is this the first place that accumulates things? You bet. If you don't think things are neat and organized around this spot, there may be some things you can try. Can you add a corner cabinet to hide things? Can you add a side table to place items on that usually end up on the floor (shoes, jackets)? If you already have a shelf or bookcase there, you might be able to add a door that hides some of this stuff.
- Out of sight. De-clutter by putting as much of the small stuff out of sight as possible. Look at all the bookcases to see if they have things other than books in them. Does it look ok? If not, try adding a door. The home store IKEA demonstrates great de-cluttering ideas and terrific use of small spaces, making rooms look bigger than they are. Consider making a field trip to Atlanta to tour these clever show spaces. If nothing else, you'll come home with a ton of creative ideas for organizing and maybe a cool gadget or two.
- Move furniture. How long have you had your furniture in the same spot? When I was a child there was nothing more exciting to me than attempting to rearrange my tiny bedroom. Half of it was the fun of trying. Give this a thought when you look around and think things are cluttered. You'll end up having fun and cleaning at the same time.
- The living room. Always an area that magnetically attracts "stuff," one way to reduce clutter is to find a nice antique trunk and use it for your coffee table. Just put everything inside it when cleaning up, and you can pull it back out when you need things.

- Hide cables and cords. Nothing produces a cluttered look more than 200 cords leaving your computer and creating a tangled mass. Take the time to disconnect everything and reassemble using ties or specially created cord covers. Even the cord snaking up the wall to your TV can be enclosed in a clever inexpensive cover. Look in your favorite home store or online.
- Finally, don't tackle everything at once. Organize one small space at a time and feel good about it. For example, begin with a cluttered closet or pantry. Focus on one small area. Have fun – go to your favorite store and buy organizing bins, shelving, or cabinets.

Don't worry about organizing the junk drawer. That's what it's for – junk. The key to making your home look un-cluttered is to have out only the things that you want others to see – like photographs or art, or books – and find places for the pet toys, tools, knick-knacks, remote controls, and jackets. Good luck!

Cleaning Gadgets: Are they worth the Money?

I am a gadget geek. I am also an "early adopter," which means I'm the person who gets in the overnight line like an idiot to buy the latest iPhone. If you combine this characteristic with an aversion to cleaning, you get a person who is trying to find gadgets to do the cleaning for them.

The three Turner Axioms that you need to know about gadgets follow (GADGET definition: "a small mechanical device or tool, esp. an ingenious or novel one"). The first is that using a "gadget" to accomplish a task will take approximately three times as long as performing the task without the gadget. Second, using a "gadget" to get something done will cost more money than not using the gadget at all. Third, by the time you unpack the "gadget" and are forced to read the manual because the on/off switch is hidden somewhere, you will have wasted four times the time that the chore would have taken had you performed it by yourself.

It was with great excitement, then, that I greeted the news of a robot to clean windows. Exactly 3.2% of the human population enjoys cleaning windows, leaving 96.8% of us doing whatever is necessary to avoid this unpleasant chore. Even many professional house cleaning people say, "We don't do windows." Don't do windows? Isn't this a part of cleaning?

Back to our nifty little robot. Called the "Winbot 7," this $350 - $400 device attaches to your window and moves with a zigzag pattern to clean dust, fingerprints, and streaks. According to Consumer Reports, this little window cleaner actually does work. It will clean glass doors and showers as well, and lends itself to those high windows that are tough to get to. The downside: you have to mount it where you want it to clean. This means you still have to get the ladder out or get into awkward positions to mount it on the outside of the windows. It would be ideal for those windows that tilt inward for cleaning though. The other downside is the price, which may drop as it becomes popular (or not). This gadget is available from Skymall.com.

The other gadget I like is for vacuuming carpet. Called the iRobot Roomba and available from Amazon.com for about $350, this pancake looking device on wheels really works. As an "early adopter" in the early 2000's I bought a Roomba and still own it. A side benefit is the entertainment factor for pets in the house. When pets hear the Roomba start, they either chase the device with enthusiasm or hide in another room. iRobot also makes a model to clean bare floors called the Scooba.

Should you purchase one of these cleaning gadgets? After understanding the important axioms in paragraph two, the best reason for getting any gadget is plain FUN. If a gadget is not fun, it's probably not really a gadget. The next time your friends are over to the house and you secretly hit the "on" button on your Roomba parked under the sofa, you will know what I mean. The look on your friends' faces as the robot wheels out from under their seats is priceless.

More Home Gadgets

Last week I talked about simple home gadgets that work well without a lot of attention. Gadgets that don't require constant fussing and fiddling are few and far between in our modern world of technology. Most gadgets that are sold as "time saving" are actually "time wasting" when you have to stop whatever you are doing and try to figure out what went wrong. Everyone has experienced that moment when you were immersed in solving one thing when you had to stop and focus on fixing something else, wondering what you got yourself in to.

So, it is with pleasure that I greet any gadget that does not need fussing and works every time you ask it to. Here are a few of my favorites.

It's almost October and Christmas is right around the corner. Are you someone who loves to put up Christmas lights? Well, maybe you do not love actually putting them up, but enjoy seeing them? Call me crazy – I love dragging them out every year and putting them in trees, on eves, around windows, and any other place where I can drag an electrical cord. And that is the rub . . . the electrical cord. My outdoor outlets during the holidays look like something out of National Lampoon's Christmas Vacation.

The gadget to solve this problem is a set of solar LED lights. They are not cheap compared to conventional lights, but my plan is to add a few strings every year and eventually the electrical cord spider webs across the lawn and driveway will go away. If you haven't converted to LED (Light Emitting Diode) Christmas lights, you should consider it. LEDs draw a fraction of the electricity of incandescents, and bulb life is much longer.

Another gadget I love is one that elegantly and reliably solves a problem you didn't know you had. And it's expensive. But it is so well made and such a pleasure to operate, I have to put it at the top of my "recommend" list. Do you have shades in windows that are hard to reach? I don't know what I was thinking when I put a panoramic window at the top of the wall where the bed is located in the master bedroom. And another large picture window looking out on to the deck and yard. Even a patio door with a window in it. During the day, it is simply lovely. But at night, when we'd like a little bit of privacy, the shades are just out of reach.

After considerable research, I discovered Serena Shades by Lutron. It took several weeks to receive these nicely packed, honeycomb cell style remote controlled shades. They were easy to install, and came with a tiny flat wireless remote that I stuck on the wall by the door. I inserted six "D" batteries (with a claimed 5-year operating life) in each shade, and hit the "down" button. All of the shades rolled down in unison with a gentle whirring sound. Wow! Then I hit the "up" button and watched all the shades retract quietly at the same time. Wow! This has now become my husband's favorite toy, and every evening as dusk approaches he runs into the bedroom to watch the shades lower. Wow! Kids and gadgets.

Time for a Robovac?

Do you love vacuuming? If you answered "Yes," you are in the minority. A little teeny, tiny minority. You might say you enjoy vacuuming, but I am guessing that you like it because in the moment that you put the vacuum away in the closet, you feel a lot better.

If you love to vacuum, don't read any further.

I myself have never been terribly excited about vacuuming. Even with the new Windtunnel vacuum with automatic retractable cord and bagless canister. You still have to set it for carpet versus bare floor, pull that cord out and repeatedly plug it into bunches of inconvenient outlets, and empty dust from the canister that sticks to everything within 10 feet.

In 2002, the company iRobot launched the Roomba® floor vacuuming robot. As a severe gadget geek, I had to have one. I saved up and ordered one through the mail. If you're geeky like me, you realize that being the first person on the block to have something is not always the best idea. The first iteration of most gadgets will have bugs and problems.

Amazingly, the Roomba that I bought did what it was told about 75% of the time, which wasn't bad. In addition to being a baby sitter for the cats, the little robot was fun to turn on with the remote when guests were visiting. As we calmly talked to our visitors, my husband would quietly press the "ON" button and the Roomba would fire up and zoom out from under the couch at their feet. This was a blast.

So here we are – almost 14 years later - have robot vacuum cleaners come of age? Maybe yes and maybe no. The July edition of Consumer Reports actually has a review of six robot vacuum cleaners. And guess what – Roomba tops the list. But these vacuums are pricey. For $675 you can own the top Roomba model; however, the lower level models on Amazon.com are half that price and the reviews are good.

Robot vacuum cleaners are now almost completely automatic. They know the difference between carpet and bare floor; they won't fall down the stairs like mine did, though they can't yet vacuum the stairs; when their battery gets low they will return to their charging base; and some models empty themselves in the base. Most come with a remote control and all can be programmed to run when you're not at home.

Downsides for even the latest robovacs include being confined to one floor; not doing well on shag or thick carpet; the ability to shatter delicate ceramics that sit on the floor; they get tangled in cords from time to time; and you will still have to empty a canister of dirt.

Upsides (if you have low pile carpet and bare floors without a lot of cords and floor hugging drapes) include help for disabled people who have trouble vacuuming; entertainment for pets and people alike; and a great pre-clean for you bare floor moppers.

Unless you are a gadget geek like me, I'd suggest waiting until Roomba can walk up and down stairs, vacuuming as it goes, empty its own dirt canister, and cook dinner.

The Latest Crowdfunding Gadgets

I know what a gadget is . . . but crowdfunding?

Like it sounds, crowdfunding is when someone starts a company to develop a new product or idea and asks a lot of people – a "crowd" to help pay for it. Before the Internet, start-ups had a tough time finding venture capitalists, angel investors, and other people willing to help them bring a product to market. Now, with social media, instant news, and online buying and selling, investors are just a click away.

This is both fun and dangerous. If you've got a terrific idea and can demonstrate it to one of the dozens of online investor sites (Kickstarter is the most recognized) then your invention may take off. But if you've overlooked something important in the design or manufacture of your product, or you've simply misjudged your market, then you and your friends will go broke.

While I refrain from investing in these sites, I'm a confirmed gadget geek, so new gadget ideas catch my interest right away. I subscribe to a "gadget newsletter" called Gizmag. At Gizmag, the editors love picking out both wild new ideas and some ideas that should have arrived already. Here are my favorite gadgets that are being crowdfunded and a few that major companies are working on.

Thermal Vision Microwave. This is something that we should have had from the beginning of microwave cooking. Why is there so much guesswork in figuring out how long something should cook? Heating leftovers can leave frozen centers or cause an overflowing mess. In the "Heat Map" microwave, an infrared lens is embedded in the ceiling of the box, streaming a thermal view of your food to a window display. As the food heats, it shows color temperatures beginning with blue and ending up white. A NASA engineer thought of this, and I hope he can sell me one tomorrow.

Whole Body Dryer. According to Tyler Overk from the Body Dryer Team, the body dryer was designed out of a need to "replace bacteria filled bathroom towels." It is capable of drying a damp adult in 30 seconds using compressed, ionized air. The base that you stand on doubles as a weight scale. If this invention is anything like the high-speed lawn mower sounding hand dryers in restrooms, I will not be buying one.

Leatherman Tread Bracelet Multi-tool. From the famous maker of pocket multi-tools, here comes a 25-function bracelet that looks somewhat normal (albeit a little clunky) when it's on your wrist but comes off to transform into socket drives, wrenches, a bottle opener, pick, hook, and flat and Phillips screwdrivers. The best part of this gadget is that you can take it on an airplane. Due this summer, this is on my have-to-have list.

The Luna Smart Mattress Cover. If you have a heated blanket on your bed like we do, the best part is slipping into it on a cold night. But if you or your significant other forgot to turn it on ahead of time, it's awful. Luna will turn the blanket on, turn lights off after you're in bed, lock the doors, monitor your temperature, breathing, and heart rate, adjust the thermostat, and turn on your coffee maker in the morning. I will buy this invention if it can also print and store money under the mattress.

What is a "Smart Home"?

Have you heard the term "smart house" or "smart home"? Every ad for security systems, video systems, lighting systems, and even some appliances say that they are "connected" and can be controlled by your smart phone and tablet. What are they talking about and should you jump in?

A "smart home" means an automated home – talked and written about in science fiction books and movies, and touted as early as the 1939 World's Fair in New York City. Automation can mean anything from programmable lights to door locks that you can unlock with your smart phone from afar. These systems include video cameras that let you know who arrived (did the kids get home from school?). The whole idea is to make living easy and simple.

However, smart homes mean electronics and gadgets, and you may have already found out that gadgets are actually not simple at all. In fact, smart home devices that claim to make things easier often make them harder and more time consuming.

An example is illustrated by one of my forays into 1990's door lock technology. The gadget geek that I am, I decided I wanted an electronic door lock. I thought it would be very cool to not use a key to get in the front door. The fancy new lock worked for 1 week and then began emitting very loud and annoying beeps at random. It refused to let me in the house. I sent it back to the manufacturer and got another one. After 2 weeks that one failed like the first one. I went back to using a key.

Now this technology is much better, but my point is that launching into full home automation should be considered with caution and research. From self-cleaning litter boxes to automatic pet feeders to light controllers to door cameras, I have tried a lot of things that ended up wasting time and breaking.

Something more worrisome is emerging, and that is "smart home hacking." If you thought your last computer virus was bad, wait until the bad guys see you through your own cameras, control your front door lock and lighting, and run up the temperature on your air conditioner. If you really want the latest tech, my advice is to wait a little longer while the manufacturers refine the security in the software.

I have decided that there are a few really good "smart home" technologies worth buying and using. These are low tech and work reliably, in order of complexity:

- Analog light timers – plug your CFL equipped lamp into this wheel timer and your lamp will light at exactly the times you choose! Completely immune to drive-by hacking and short term power outages.
- Occupancy Sensor Switches – replace a traditional light switch with a motion detector. The light comes on when you enter the room, and turns off when you leave. Wow!

- Simple Pushbutton (not Wi-Fi) door lock – no key needed – punch in your 4 or 5-digit code, you're in. I've been using one for six years.
- A top end programmable thermostat for your heating and cooling – program and forget.

If you decide to jump in to smart home devices now, remember that anything connected to your internet wireless router ("Wi-Fi") is potentially hackable, so make sure you have a robust password, making it tougher to crack. Have fun.

The Argument for a Dumb Home

Do you wish you had a smart home? Perhaps you already have one. A smart home is one where the lights know when you are coming home, the thermostat has adjusted the temperature for your arrival, the window shades arrange themselves for evening, the Wi-Fi crockpot has your dinner ready, and the refrigerator tells you what has been used up and needs to be replenished (and places the order). The cameras send you pictures of delivery people and you can talk to your dog or cat while watching them through cameras. Your garage door opens as you approach the house and your door locks click open as you near the door.

Even though I am a gadget freak, this level of smartness may be too much. Here's why.

First, Wi-Fi devices continue to rely on different protocols and controllers are still time consuming to set up. The old X-10 controls from the 1990s were simple compared to INSTEON, ZigBee, and Z-Wave. Choosing a home hub is complicated. Asking all of the devices to play nicely with each other may not work. As I have found out myself through experience, at any given time there will be at least one thing that is not working or communicating and has to be adjusted, reset, or fixed.

Dozens of manufacturers are making smart home products. These include Iris, SmartThings, WeMo, Nest, Wink, Nexia, Vera, and Harmony. Varying levels of ease accompany these new products. Nearly all of the reviews that I've read talk about problems. While it may be simple to get your cameras working over WIFI, when you start adding other products you may discover that some gadgets don't work.

Second, some bad news came out of the latest Black Hat electronics hacker conference earlier this month. It's probably good that hackers assemble to best one another on breaking into things, but bad that there are people who make a living doing this.

Attendees at this year's conference could watch hackers remotely hijacking consumer WIFI accounts (simulations) and locking/unlocking cars remotely, opening electronic safes, turning off a "smart refrigerator," turning up a home WIFI connected thermostat, changing settings on "smart scales," and commandeering baby monitors, laptop cameras, and garage doors. As manufacturers are rushing to put wireless connectivity in to electronics devices, hackers are rushing to show that the products are designed with a disregard for the most basic security principles.

Having said all that, should you worry?

Probably not. Being fortunate enough to live here in the mountains also carries a measure of protection. It is far more likely that someone in New York City will get their home or car hacked into than someone here. In the suburbs and in the city, it's easy to park in front of someone's house to run software to crack passwords. If someone did that here we'd be walking out to them and saying, "May I help you?"

My advice is two-fold. First, consider adding technology that's already been out for a while, like Logitech or Nest cameras. Consider waiting on a controller until there's more security and more interoperability. Second, make sure that you have a good password on your WIFI system. No WIFI? You are the most protected of all.

The confirmed gadget geek that I am will continue to embrace and experiment with the latest smart gadgets, but I'm not sure that the integrated "smart home" is ready for prime time. If you love these gadgets, by all means go ahead and invest in them. Either experiment like I am doing, or find a professional who can give you sensible advice so that your money is not wasted.

Remember that your WIFI network should have a very rigorous password, even here in the lovely mountains of Clay County where many of us don't even think to lock our doors. If you do have an assortment of smart gadgets connected to your WIFI network, you want it all to be secure. The internet is a very big place and hackers halfway across the world would love nothing better than to get in to your personal network.

The chances of finding a home buyer who loves all this technology is still statistically low. So, if you invest in these systems, do it for yourself. Realize that in the future you may need to upgrade the gadgets. And remember that the top "smart" gadgets that homebuyers like are pushbutton door locks and security cameras.

Home Security Checklist

The first "bite" of fall weather has arrived. The holidays are right around the corner. This is a perfect time to review your home security as you make your lists preparing for gift giving and the social schedule. It's easy to forget about security as we're bustling about with other activities, but the less honest 2% of the population has a contingent here in the mountains too. The following checklist will help you stay safe, along with your possessions, all year long.

- Play burglar. Leave your home as you would if you were driving down the road to the grocery store. Before you go, take 15 minutes to walk completely around the outside of your house. If you wanted to get in the house without a key, could you? Look at windows, doors, access to a second story, and the garage entry. Make note of weaknesses, and consider how you might make it more difficult for someone to break in. It is true that someone who is determined to enter your home will find a way. But my philosophy is to make it more difficult and discouraging for the casual burglar.

- Talk to your neighbors. Our mountain homes are not like the cookie cutter developments of Florida with 10 foot easements. We may not be able to see our neighbor's house from ours. If you all agree to keep an eye out for each other's property, it will be tougher for a burglar to remain unseen.
- Lock your door. There are folks who do not, and will not, lock their doors. The chances that someone from that 2% aberrant group will come through your door, or break in while you are away, is small. But the chance is there, and we do read about it in the news. My philosophy is, "better safe than sorry."
- Put motion detection lights at entrances. Replace your entry light with a sensor light. It will come on and turn off automatically, making the home look active. It will usually frighten a casual burglar.
- Don't put the key under the mat. Yes, I have done this too. Find a more creative place for the key. Velcro it under something that only you and your family know about away from the entry, or under a special rock in the garden.
- Don't leave your home completely dark. Buy an analog timer to plug a few table lamps into. Some timers have a variable setting so that they do not come on or go off at exactly the same time.
- Consider a video cam. If you have Internet service with WIFI in your home, you can purchase a wireless camera that sits in any room or window, and sends an alert to your smartphone if it sees something. If someone breaks in, you'll know it instantaneously, and have a picture of the event. A great unit is called "Dropcam" and is available locally or from Amazon for about $150.

These are just a few of the things that you can do to stay safe. Awareness and forethought will go a long way. Add your own ideas to the list and review it periodically.

Keeping Your Home Safe from Intruders

Around the holidays we see extremes – joy and happiness around this special time of year enjoying our families, friends, and the celebration of the season – but also notice that thefts and break-ins seem to rise over the holidays. In last week's Progress, we read about some of these events and were surprised, because we tend to think that our special mountain communities are above this universal problem. While residential break-ins do not take over our news headlines year-round, we should still take proactive action so that we are not part of the next news story.

As a home inspector, I find security issues all of the time with the homes that I inspect. It is only because break-ins are relatively rare that these homeowners have not experienced a problem. Here are some things you can do to keep your home safe from intruders without a lot of expense.

Take the Sherriff's advice – consider keeping all of your entry doors to your home locked, including the garage. You may have grown up in this area not worrying about this, but unfortunately times have changed. It only takes that 2% of the population who wish to take our belongings or do us harm to create a problem. Burglars love garages – if they can get in to your garage, then they can spend their time breaking into your house. Consider putting a simple motion alarm in the garage – they will think the whole house is alarmed and likely leave.

Burglars know your hiding places for spare keys. Unless you can get really creative, don't hide a key outside. As a home inspector I have seen "hidden" keys about 25% of the time.

Where do you store ladders? If they are not locked in your garage or in a shed, they make convenient tools for upstairs window access. Make sure second story windows are locked. Another favorite entry is through sliding glass doors, where thieves can simply lift the doors off the track. To prevent this, place a dowel cut to fit into the inside track, or place a pin through the center frame that can be removed easily when you want to open the sliders.

Take a good look at your door lock hardware. Is it secure? Make sure your doors can't be opened with a flat bladed screwdriver – the best way is to install a deadbolt lock in addition to the knob lock.

Should you get an alarm system? This is a personal decision and depends on many factors. Alarm systems can be simple or complex, expensive or inexpensive – if you are considering one, I recommend getting several free quotes from reputable companies. You can also set up video monitoring yourself with alerts delivered to your cell phone. These options may be a good idea if you are away from your home a great deal or if it's a second home getaway. Another trick – get the alarm stickers for your windows . . . even if you don't have an alarm. Thieves won't want to take the chance and will move on.

How to Add Video Security to Your Home

Several Progress readers wrote this week to ask me about video security for their homes. One reader said she has a cabin that she leaves vacant for the winter and would like to know if things are ok while she is away. Another reader said that he assumed adding video would be very difficult and complicated. Since you may have similar questions, I thought I'd provide some answers.

Do you have high speed internet service at your home? If the answer is yes, then setting up video monitoring is straightforward. If the answer is no, then you will not be able to monitor cameras from your smartphone or a remote computer, but you can record what happens on the cameras for review later.

The following questions and answers assume that you do have internet service and either have, or can hook up, a WI-FI network in your home. This is simply a box (called a "router") that connects to your internet service and allows devices to communicate wirelessly inside your home.

Do you already have a security system that is monitored by an alarm company? If so, then it may be a good idea to ask them about video. Most alarm companies can place cameras in your home as part of their service. However, they may want to charge you more money. If this means peace of mind, then it may be a good tradeoff.

Are you building a new home? If the answer is yes, and you have not started the electrical wiring, you're in luck. Ask your electrician to run camera feeds. Purchase a wired camera system locally or from Amazon.com. All you need to do is decide how many cameras you want. Four to eight cameras should be plenty. A four camera system can run less than $400. You can run the feeds yourself if you're handy. Wired systems offer excellent picture quality and are WI-FI capable for remote monitoring.

Do you want to add a camera system to an existing home but can't run wires through the walls? If the answer is yes, there is a very simple way to get video that is good quality and not expensive or complicated to set up. I recommend either NestCam or Logitech Alert. These are available locally and you can research them online. Both have cameras that plug into the nearest power outlet and communicate over your WI-FI network. Dropcam stores video out on the "cloud" for a fee, and Logitech stores footage on the camera and on your PC for free.

Do you want to monitor your vacation cabin while you are in Florida? Use a smartphone or a computer to observe, and even listen, to your home. You can use any of the above solutions to monitor your property in your absence. The catch? You will need to leave the electric on, and continue to pay for your monthly internet service.

I offer one caveat to all of this excitement: make sure that you choose a good password for your internet wireless router. If someone hacks in to your network, they will be able to see what you see!

Home Going Away Checklist

Leaving home? Whether it's a vacation, visiting family and friends, or going to another home – we sometimes take for granted that all will be well when we return.

I have heard enough scary stories about folks coming back to water flowing out the front door to a picture window punched out from a tree branch. How likely is it for these things to happen? Not high, but always in the realm of possibility.

If you're going away for a few days to months at a time, I would follow a checklist to keep your home safe and sound. Have you ever left the house on vacation and then 10 miles down the road said, "Honey, did you lock the front door?" Use the following tips to provide peace of mind while you are away.

Winterize or not? Winterizing a home is complex and should be done by a professional. The procedure involves removing all of the water from the appliances, fixtures, and plumbing so that freezing temperatures do not produce damage in these systems. This is done when the intent is to turn off all systems, including the electricity, for long periods of time (usually 6 months to years). As an inspector, I found damage in winterized homes because either the winterizing or de-winterizing was done wrong. In many cases, not all the water was removed from the pipes and they burst. In other cases, I saw extensive damage from mold due to uncontrolled humidity. The temperature cycles inside a winterized home can shorten the life of materials inside as well as warp doors and floors.

I would only winterize your home if you absolutely must turn off the electricity. My opinion is that you would be better off to keep the heat on and find a reliable person to either live in the home while you are away or pay for a house check service.

Now that we've decided to keep the heat (or air) on in the house, here is your checklist.

- Turn off the water supply coming in to the house. Flush each toilet to remove pressure.
- Turn off the electricity to the water heater. This will be a well-marked breaker switch in your electric panel.
- Unplug your computers, TVs, and electronics. This will save them from power spikes and reduce the cost to keep them on standby mode.
- Set the thermostat on 55 degrees in the winter and 82 in the summer. If you are nervous about the heat (or cool) staying on, consider installing a WI-FI connected thermostat. If you keep your router plugged in to maintain the wireless signal, you can check on the temperature, or change it, from your smartphone or computer.
- Connect several lights to analog timers that will make your home look lived in.
- Make sure no hoses are connected to outside bibs. Even if your faucets are "frostproof" versions (that do not hold water in the line), place an insulating cover on them to prevent the line from freezing where it goes through the siding.
- Finally, I recommend that you have someone trustworthy checking on your home from time to time, especially if you are going to be away for a long time.

Don't forget to lock the doors!

Is it Time for a Home Electronic Assistant?

Over the years, I've written about kitchen gadgets, vacuuming gadgets, garage gadgets, security gadgets, and window cleaning gadgets. I am a gadget magnet. I'm not sure whether they attract me, or I attract them, but I revel in finding out about some new electronic do-dad and testing it. I should go to work for Consumer Reports Magazine.

As the "digital assistant" craze took hold through our smartphones years ago, I wasn't very impressed. First of all, you had to have a Wi-Fi connection for the digital assistants to answer intelligently, the assistant needed to know your location, and even in peak form these assistants didn't have very good answers. I admit they do have a sense of humor, but when you are searching for something important and your digital assistant is impertinent, I have a tendency to become annoyed. There are already enough stressful situations in life. The ads made it sound wonderful, until you actually try dialing someone by voice as you are jogging the trail and get someone you don't know.

The latest development are kitchen counter or side table assistants. These are medium sized voice activated speakers that also need a Wi-Fi connection to the internet. The most well-known units are the Amazon Echo and Google Home. These assistants have come a long way in terms of technology and can actually be useful and fun. Things you can ask them include, "What is the half time score of the NC State Duke game?" and it will actually understand that you mean the current college basketball game between these two teams and give you an accurate answer. Or, you can say, "Google, play music from 1998," or "Alexa, order me an eighteen once box of liquid Tide HE."

If you are thinking about buying a digital assistant for your home, here are the pros and cons.

Advantages include instant smart (and some funny) answers to most of your questions; they can play music on command, including from your phone or from a free service like Pandora; and when you are cooking they can reference a recipe or serve as your kitchen scientist, as in "How many milliliters are in an ounce (29.57)?

Disadvantages are that you have to have a Wi-Fi (wireless internet) connection that is not turtle speed; setup requires a smartphone app, and do you really want one more person talking to you when you get home from work?

Some folks have thought it might be handy to have one of these assistants as a "friend" for their elderly parent that might not only serve as a companion that they could talk to, but also be able to call for help. In principle, this is a great idea. But until we can get past the smartphone setup and requirement for Wi-Fi, this is still somewhat impractical.

If you are undeterred by these drawbacks, which assistant should you get? If you just want the smartest answers plus some home automation, choose Google Home. If you order things constantly from Amazon and use their Prime Music service, then the Amazon Echo is the way to go.

Or, you can be like me and arrive home to the sound of the creek and a gentle wafting breeze through the trees.

7

Heating, Cooling, and Venting

How to Prevent Early HVAC Failures

Think about your automobile a moment. As you are getting into your car, you're thinking about the work day, your family, a recent crisis, or some other consuming issue. The car starts, you put it in gear, you drive off. You get to where you are going, turn it off, and start the day.

We don't think about the car. Before we get in the car, do we say to ourselves, "Gee, I better check the tire pressures. I should talk a walk around the car to see if there are any new dings, or items falling off, or a flat". No, we turn the key and drive off. We are lucky that our cars are so reliable!

What does this have to do with HVAC, or the heating and cooling systems in our home? Here's the connection: we don't pay any more attention to our heating and cooling equipment than we do to our cars. Cars are built to take this neglect, and survive it pretty well. But HVAC equipment is not built to run without regular maintenance. So, as a home inspector, when people ask me to tell them how their air conditioner or furnace is, I can tell them about its current state – but can't guarantee that the unit will keep working for them, because most of the time folks think it does not need regular maintenance.

This misconception ends up costing consumers a lot of money, since 12-14% of your income every year is being spent on electricity or propane for these beasts.

As much as you may not want to spend the money to have a technician check your heating and cooling equipment, you will find that this regular servicing – checking refrigerant pressures, balancing air flow, and making sure ductwork is intact – ends up saving you money later in the form of longer service and lower energy bills.

There is one thing, though, that YOU can do to prevent early equipment failure. I see it all the time on my inspections. You know what it is. You got it! Dirty furnace/air conditioning filters. Not changing filters often enough is the cause of more air conditioner, heat pump, and furnace failures than any other item. In addition, leaving dirty filters in place will end up coating your entire ducting system with dust, dirt, and allergens. This gunk will then cover the fan motor in your air handler and cause it to overheat. The high temperature switch cuts the motor off, and you then have some expensive energy consuming short cycling going on . . . until it fails . . .you get the picture.

Here's what to do. Look around and determine where ALL the filters are. You might have just one, usually at the air handler itself, or you could have many throughout the house. They might be at floor level, and some might even be in the ceiling. I have never figured that one out – why would a builder put a filter in the ceiling? Ok, it is more efficient – but It is a big inconvenience for homeowners – no wonder the filters stay dirty! You might also find a filter at the air handler in the attic – also inconvenient, but important. Your home may have a single or double set of <u>electronic</u> air filters for your HVAC system. These are great filters, but you should read the instructions for them carefully before cleaning, or they could be damaged. If you get bi-annual tech inspections, you can ask them to show you how to check and clean them.

Start by changing your filters every 2 months. If they look clean, go to 3 months. If dirty, go to one month. Once you've figured out the schedule, put this maintenance task on your calendar.

I may get disagreement on this last piece of advice, but I'm sticking to it: buy the cheapest filters you can (make sure they are the right size) and buy a can of filter spray at your favorite home improvement store. Spraying each filter before you install it will help filter allergens, improve the trapping of dirt, and improve performance, costing much less than what you'd spend on high end filters.

Space Heaters: How to Choose

In the last column I talked about not getting suckered into buying a particular brand and type of space heater because of "special mail offers." I always turn to the reviews by independent testing labs to see what real people, not ad writers, are saying. Now that you know you can buy 1500 watts of heating power in a basic heater for $175 or less, how do know what to choose?

The good news is that space heaters are efficient. 100% of the electricity that goes in to a space heater comes out as heat. The amount of heat generated is rated in "watts" which is a way to determine how much electricity the heater is using. And it's easy to figure out how many watts you need: estimate 10 watts per square foot and multiply the square footage of the room by 10. For example, a room with 144 square feet will require 1,440 watts. When we say square feet, we are just saying measure your room with a tape measure – length and width – and multiply the 2 numbers.

Once you know how many watts you need, you can calculate what it will cost over a year by multiplying the watts times how many hours you have it on times the rate per kilowatt-hour that the electric company charges. Here's an example. If you have a 1,500-watt heater and you leave it on 24 hours a day at a rate of 9 cents a kilowatt-hour, the cost would be **$3.24 a day** (1,500 x 24 ÷ 1,000 x $0.09). Over a month that would amount to $97.20; a year it would be $1,166.40.

How can a space heater help save on energy costs? While the costs in the last paragraph are indeed high, you can still use a space heater and save a little money every month by doing the following.

- Only run the heater when you are in the room.
- Buy an appliance timer to use with your heater if you think you'll forget to turn it off.
- Lower the thermostat in the rest of the house. Obviously if there are a lot of people in the house, or you're going from room to room, this strategy will not be comfortable.

Although you will hear words like radiant, oil filled, infrared, ceramic, quartz halogen, and convection in heater descriptions, there are really only 2 types of heaters: radiant and convection, plus combinations of these two. Radiant simply means a strong heating element, like a halogen bulb, that radiates heat. These heaters work well for spot heating. Convection heaters use wire or ribbon elements and a powerful fan to move the heat through the room. They are better for heating large areas. A combination unit, such as a radiator type, uses water or oil circulating in metal fins to produce heat. They don't get as hot and give off a constant, steady heat.

Any of these heater types are available for less than $200 unless they have fancy accessories like remote controls. Do your homework, and you'll save money and stay warm too. Just what we need as winter approaches.

Choosing a Space Heater: Don't Get Taken

Fresh, cool air is drifting through the color of the autumn trees and some of us are turning the thermostat to "Heat." Smaller homes and vacation cabins often do not have a furnace, so we rely on the fireplace, a wood stove, or electric heat baseboards to keep the house livable in the winter time. Even in homes with central heating, a small "space heater" can be a handy addition when we only need heat in one chilly area.

This week in the mail I received an important looking envelope marked "Affordable Home Heating Initiative" in the upper left hand corner. In a box below this it said, "Prepared exclusively for a limited number of residential users in North Carolina. Approved participants guaranteed a discount credit up to $229. PLEASE REVIEW AND RESPOND IMMEDIATELY."

Well, that got my attention. What in the world could this be? I opened the envelope to find an "Authorized Discount Credit Claim Form" for . . .. a space heater! As I read the amazing claims for this "21st century technology," I started laughing. Apparently, if you turn your central heat down to 55 degrees and then run the heater only in the room where you are, and then unplug it and drag it to the next room you're in, you will save 50% on your heating bill. I guess you'll need to walk around in an Eskimo parka.

I read on. The price, after my "special discount," is only $247, if I am one of the chosen residents able to take advantage of this fabulous offer. Otherwise, I pay $476. I put down these amazing materials and turned to the internet. First, I go to Amazon.com and see that I can buy variations of this same heater for $268, not $476. Next, I go to Consumer Reports (www.consumerreports.org). Consumer Reports does not accept advertising, and I like their objectivity and thorough testing. Turns out that not only is this "special offer" heater not anywhere near the top of the ratings, it is more expensive than some of the top-rated models. One of the heaters at the top carries a price tag of $40.

Space heaters are handy and convenient, but don't count on one to heat your home exclusively, or to save much on your heating bill. Electric heaters, even those claiming "infrared technology" are expensive to operate compared to central heat systems such as heat pumps and propane furnaces.

With that said, all of us have used and enjoyed space heaters, and they can be just wonderful in those areas where it's a little cold and you need a boost. Under your desk, in the workshop, and a cold bathroom become cozy places with space heaters. So, if you are thinking about buying a space heater, hit the internet yourself (or ask a teenager to help you if this is not your favorite thing) and learn everything you can. Reading the reviews on Amazon.com can tell you a lot too.

If you're not inclined to do all this homework yourself, wait until the next column, where I'll explain the different types of space heaters, provide size and type recommendations, and give you some tips and tricks for those elusive energy savings.

Science Fiction? Latest Technologies in HVAC

If you are building a home, remodeling a home, or thinking of changing out your old heating and cooling equipment for new, take a quick look at new technologies that make these systems more efficient and green friendly. From saving money to improving indoor air quality, the latest HVAC systems have come a long way. Heating and cooling accounts for over 50% of the energy use in a typical U.S. home, making it the largest energy expense that we have.

HVAC improvements that are available now include variable speed compressors that yield much better efficiency, air exchange systems that exhaust stale air and deliver fresh and clean, conditioned, air, specialized zoning to deliver different temperatures to individual rooms, sub slab zoned radiant heat, and controls that can be programmed and run through the internet.

Ductless heat and air systems, known as "mini-splits," have become quieter and more efficient through advances in compressor and fan technology. These systems deliver warm or cool air directly into rooms in your home, instead of being routed through ducts. They are a cost-effective solution to replace inefficient baseboard electric heating and window air conditioners in older homes. They are also used in new construction, home additions, multi-family (condo or apartment) housing, and to improve comfort in poorly heated or cooled rooms.

You can also choose a "Mini-duct HVAC" system. These are relatively new and have small (2-inch flexible tubing), high velocity air delivery to rooms and the system is easy to install and retrofit into existing construction. The ducts are almost invisible, and the system is quiet. The high pressure helps dehumidify the air better than traditional systems. Mini ducts also help distribute the air better inside rooms and yield high efficiency.

Geothermal HVAC is heat pump technology (does not use fuel) that heats and cools by running refrigerant through specialized pipes in the ground. The ground maintains about 55 degrees year-round, so the system does not have to work as hard as it does when using hot air in the summer and cold air in the winter.

Some of the latest experiments underway sound like science fiction:

New refrigerants. A variety of new chemicals, including plain old saline, could be used to reduce costs and improve efficiency in compressors. These new compounds would be recyclable and safer than the current products.

Solar Air Conditioning – How about using solar energy to power a cooling and heating system? Doesn't this sound space age? Developers are working out the pesky details, like how to store energy for night time running.

Sound Waves – Cooling with sound waves? Scientists are working on this technology. Who would have thought we could heat food with microwaves? This new technology will also appear magical at first.

Whatever system you investigate, be sure to search out a competent installer. If you're building a new home, I recommend you engage the consulting services of an experienced HVAC design engineer. You will get that money back many times over in an efficient installation.

Heating and Air Conditioning Tips and Tricks

So, as a home inspector, when people ask me to tell them how their air conditioner or furnace is, I can tell them about its current state – but can't guarantee that the unit will keep working for them, because most of the time folks think it does not need regular maintenance. This misconception ends up costing consumers a lot of money, since 12-14% of your income every year is being spent on electricity or propane for these beasts.

As much as you may not want to spend the money to have a technician check your heating and cooling equipment, you will find that this regular servicing – checking refrigerant pressures, balancing air flow, and making sure ductwork is intact – ends up saving you money later in the form of longer service and lower energy bills.

There is one thing, though, that YOU can do to prevent early equipment failure. I see it all the time on my inspections. You know what it is. You got it! Dirty furnace/air conditioning filters. Not changing filters often enough is the cause of more air conditioner, heat pump, and furnace failures than any other item. In addition, leaving dirty filters in place will end up coating your entire ducting system with dust, dirt, and allergens. This gunk will then cover the fan motor in your air handler and cause it to overheat. The high temperature switch cuts the motor off, and you then have some expensive energy consuming short cycling going on . . . until it fails . . .you get the picture.

Here's what to do. Look around and determine where ALL the filters are. You might have just one, usually at the air handler itself, or you could have many throughout the house. They might be at floor level, and some might even be in the ceiling. I have never figured that one out – why would a builder put a filter in the ceiling? Ok, it is more efficient – but It is a big inconvenience for homeowners – no wonder the filters stay dirty! You might also find a filter at the air handler in the attic – also inconvenient, but important. Your home may have a single or double set of <u>electronic</u> air filters for your HVAC system. These are great filters, but you should read the instructions for them carefully before cleaning, or they could be damaged. If you get bi-annual tech inspections, you can ask them to show you how to check and clean them.

Start by changing your filters every 2 months. If they look clean, go to 3 months. If dirty, go to one month. Once you've figured out the schedule, put this maintenance task on your calendar.

I may get disagreement on this last piece of advice, but I'm sticking to it: buy the cheapest filters you can (make sure they are the right size) and buy a can of filter spray at your favorite home improvement store. Spraying each filter before you install it will help filter allergens, improve the trapping of dirt, and improve performance, costing much less than what you'd spend on high end filters.

How to Buy an Air Conditioner

Is it summertime yet? I think it is! Following the oddly warm winter, we now have an oddly hotter summer. The mountains do give us some moderation over the temperatures that the rest of the country is seeing; even so, most of us rely on air conditioning for comfort during the really hot days.

The hottest days seem to be when our AC decides to quit. In that moment when we realize the coolness is gone, we panic and will do just about anything to get the darn thing working again. Here's a brief guide to picking out a new window or roll around air conditioner if your unit bites the dust.

Air conditioner capacity is measured in "BTU," or "British Thermal Units." 12,000 BTUs equals "1 Ton" of cooling. 1 Ton of cooling represents the amount of heat required to melt 1 ton of ice in 24 hours. Is this making any sense? I didn't think so either. Those Brits are something. Anyway, for our discussion, we'll just say that "cooling power" depends on the size (BTU) of the unit you are considering.

The biggest mistake people make is buying too much BTU, or cooling power, for the space that they have. Do not think that "more

(BTU) is better." Too much cooling power will leave you cold and damp. This happens when the unit only runs for a short period of time (called "short cycling") and does not run long enough to dehumidify the air. This is not only uncomfortable, it also puts more stress on the air conditioner and shortens its life. The AC experts all say it is better to undersize your unit, although getting a unit that is too small will not have the oomph to cool you on the really hot days.

How do you choose the right size? Simply measure the room's interior length and width and multiply the two to get the square footage. Add 10% to this number if the room gets a lot of sun, and another 5% if you have ceilings higher than 8 feet. If your room is 100 to 300 square feet, pick a unit that is rated at 5,000 to 6,500 BTU. 250 – 400 SF – get 7,000 to 8,000 BTU; and if your room is 400 to 700 SF, choose a unit with 10,000 to 12,500 BTU. Rooms larger than this are candidates for central AC, since window units are not designed to cool really large rooms.

Hate putting a unit in your window? For a little more money, you can purchase a small air conditioner (some also heat) that rolls around inside your room. You'll need to run a flexible duct out the wall, however, to exhaust air and condensate. We have one of these at our house to cool the bedroom at night, saving us from cooling everything else in the house while we are sleeping.

If you need advice on the most reliable air conditioning units along with pricing, you can go online for information or consult Consumer Reports Magazine. Consumer Reports does not accept advertising and offers expert, impartial reviews of products, and even tells you where to get the best deal. Stay cool!

Winter Heating Safety Tips

It seems like winter arrived in the mountains suddenly, as if someone flipped the "winter" switch to "on." Switching from air conditioning to heating can present surprises, like when it doesn't work and it's 22 degrees outside.

Our expectation is that the heating equipment will work perfectly the first time in over 7 or 8 month of being idle when we hit that thermostat, but this isn't always the case. It's sort of like driving on tires when we've not checked the air pressure - we don't really notice diminished performance over time. Or, we may find on the first cold day that the heat does not work at all. This is one of the "things your mother told you but you didn't want to hear" sayings: regular and preventative maintenance pays off over time. You can go year to year and the appliances continue to work, but if you have a plan to service them regularly, they are less likely to surprise you with a malfunction.

Here's how to keep your heating appliances working and keep you safe.

Fireplace. Most of us throw the wood in there and light it. That's it. Simple. But wait! Before your next fire, take a look at a couple of things. First, is the damper working properly? Is it closing all the way? Opening all the way? If it's not, you're sending warm/cold air up the flue when you're not using the fireplace, and when you are using it, it may not be venting properly. Take your fingers and press at the rear of the firebrick inside the fireplace - is there any give, or any loose brickwork? This is important, as loose brickwork can cause the area behind it to receive more heat than it should and cause damage. Open the damper and look up into the flue, or use a digital camera and point it up the flue and take a picture. Is there a buildup of creosote, a fire byproduct? If you discover any problems. call a professional chimney sweep for an inspection and a cleaning. Creosote buildup can cause a fire where you do not want it - inside the chimney.

Gas Logs. At the beginning of the heating season, have your gas company inspect the gas lines and valves. Try to negotiate this service into the purchase of your propane. As a home inspector I always look at gas lines that are visible and do the "sniff" test - get your nose right up to the valve and see if you can detect any sulfur, or "rotten egg" smell. Propane is a heavier than air, odorless gas that has the smell added to it so that we can recognize a leak. I check the outside storage tank valve too. About 8% of the time I find a leaking line or valve. This is serious - a spark or flame can cause propane in the air to explode. If you do think you have a leak, call your propane company immediately and exit your home; do not turn any electrical light switches on or off as you leave. They can generate a spark that could ignite the gas.

Gas Furnace. If you have a gas furnace, follow the tips above for leak detection. What I'm about to say here now also applies to gas log and wood burning fireplaces: these appliances can produce small amounts of carbon monoxide if there is a leak, or if venting is not adequate. Both propane gas and carbon monoxide gas settle to the floor, so I recommend that you add a carbon monoxide detector near floor level if you don't have them installed in your home now. Get them at any home improvement store and follow the directions for installation.

Safety Tips for the Holidays: Heating

Carbon monoxide leaks and exposure can happen any time of year, but increase during the winter when we are using gas and wood heating appliances. If you don't have carbon monoxide detectors in your home, you should make a Christmas present to yourself and your family and buy one for each area where you have propane and wood burning fireplaces, appliances, or stoves. Yes, burning wood can also generate CO gas if there is incomplete combustion. Detectors are available locally and online for less than $25 each.

Regular service checks are important for your appliances. This is one of the "things your mother told you but you didn't want to hear" - regular and preventative maintenance heads off problems. I know you're saying "It turns on and it's running fine, why worry?" You should worry because systems degrade and lose efficiency over time. Here are some simple inspections you can do yourself.

Fireplace and wood stove. Most of us throw the wood in and light it. Simple. But wait! Before your next fire, take a look at a couple of things.

First, is the damper working properly? Is it closing all the way? Opening all the way? If it's not, you're sending warm air up the flue when you're not using the fireplace, and when you are using it, it may not be venting properly. Take your fingers and press at the rear of the firebrick inside the fireplace - is there any give, or any loose brickwork? This is important, as loose brickwork can cause the area behind it to receive more heat than it should and cause damage. Open the damper and look up into the flue, or use a digital camera and point it up the flue and take a picture. Is there a buildup of creosote, a fire byproduct? If you discover any problems, call a professional chimney sweep for an inspection and a cleaning. Creosote buildup can cause a fire where you do not want it - inside the chimney. Make the same checks on your stove, and look for areas on vents and ductwork that might be rusting or damaged.

Gas appliances. At the beginning of the heating season, have your gas company inspect the gas lines and valves. Try to negotiate this service into the purchase of your propane. As a home inspector I always look at gas lines that are visible and do the "sniff" test - get your nose right up to the valve and see if you can detect any sulfur, or "rotten egg" smell. Propane is a heavier than air, odorless gas that has the smell added to it so that we can recognize a leak. Check the outside storage tank valve too. About 8% of the time I find a leaking line or valve. This is serious - a spark or flame can cause propane in the air to explode. If you think you have a leak, call your propane company immediately and exit your home. Don't turn any electrical light switches on or off as you leave. They can generate a spark that could ignite the gas.

Now that you've done your chores, enjoy the holidays.

How Clean is Your Indoor Air?

If you suffer from allergies, you probably assume that they are from outdoor pollutants. You may have been told to stay inside and close the windows to reduce your exposure to these irritants. Although many of these irritants do come from outdoor plant materials, did you know that the air inside your home is more than twice as "polluted" as the outside air?

Indoor air can contain a multitude of harmful substances, from dust contents to byproducts of cleaners, plastics, and fabric fibers. These hidden residues can aggravate an allergy sufferer and blur the line between outdoor allergens and indoor allergens. What can you do to improve the air inside your home?

Start with dust. Dust appears harmless because of its uniform layered appearance. But do not be fooled; the EPA analyzed dust particles in a study of indoor air quality and found everything from mold spores to PBDEs (chemicals used in flame retardant fabrics) to pet waste products and the miscellaneous stuff that comes in on the bottom of your shoes. Better to not get your microscope out and look!

Is dusting your favorite activity? Not! Nor is it mine. But we have to get that dust up, and do it regularly, to have cleaner air inside our homes. Statistics say we spend 90% of our time indoors. This is a lot of time breathing inside air. Follow these tips to immediately improve the quality of your indoor air.

Vacuum using a model that can use a HEPA (high efficiency particulate air) filter. These filters have much finer mesh that trap the majority of contaminates and compounds. Vacuums that can handle these filters are more expensive, because they require more air pressure to operate, but are worth every cent. Why? Because vacuuming without filtration spreads contaminates rather than trapping them. Yes – you figured it out – they are worse than not vacuuming at all. The carpet may LOOK nice but all you have done is redistribute the dust and everything in it.

Make sure your furnace and your air conditioner have filters and check and change them often. What this will do is greatly reduce the amount of dust that you end up with over time, and trap many of the nasty items we listed above. When you go into your favorite home supply store, you will see a range of air filters. Choose the filter with the highest MERV (minimum efficiency reporting value) rating you can afford. Alternatively, you can purchase a can of filter spray, which will improve the particulate capture on the cheaper filters. Can you get a HEPA filter for your air conditioner? No, you cannot – because they are so restrictive. However, you can add a separate media unit or electronic filter to your central air conditioner. Don't bother with the "ionizing" air filters that were the rage years ago. These standalone plug-in units actually may generate ozone, which is a chemical irritant.

If your home is tightly constructed and does not have a system for bringing in fresh air from the outside (sometimes called HRV or ERV), consider having an HVAC contractor quote installing one. These units are efficient and guarantee a certain level of filtered fresh air exchange in your home.

Open the windows! Unless you have ragweed growing outside your window, letting outside air in will help clear some of the suspended contaminates in your home.

Reducing Indoor Air Pollution

In the last column, we talked about ways to clean the air inside our homes. The clean-up of indoor air starts with dust, and we covered ways to reduce dust buildup in our homes and apartments. Here are more tips and tricks in reducing indoor pollution.

Chemical exposure. Typically, we think of "chemicals" as being dangerous and in the province of scientific laboratories. However, "chemicals" are good and bad, and range from critical-to-life compounds (good) to VOCs (volatile organic compounds – bad) such as benzene and formaldehyde. In our homes chemicals are constantly outgassing from furniture, carpet, wall coverings, and building materials, and are released through disinfectants, cleansers, and air fresheners. What to do? Consider buying used furniture, where the chemicals have already dispersed, or we can go to IKEA or other furniture stores locally that have committed to VOC low materials in construction. If we're building a new home, we can specify low VOC carpets, paints, and a variety of other "green" building materials. Cotton and wool are natural materials to look for in upholstery.

Cleaners. Use strong cleaning chemicals sparingly. Try diluting them with water – they will last longer and be safer. "Green" cleaners are also available that do not contain harsh chemicals.

Pets and guests. Sorry to place these two together, but they tend to bring in similar "stuff" from the outside. From pet dander and parasites to the material that rides in on shoes and baggage, our homes are host to a variety of irritants. Consider placing area carpets in guest rooms and encouraging guests to keep doors closed. Shake out (outside) or vacuum these carpets regularly. Consider not allowing pets on your furniture or into your bed. This may not be possible if Fido or Fluffy is a spoiled bed mate already (like mine) but it will reduce the amount of dander, mites, and other assorted debris from their daily adventures collecting in these areas.

Humidity. To improve your comfort level inside your home and reduce allergic type reactions to dry air, go to your favorite home supply store and buy a hygrometer. These devices are inexpensive and will read the amount of water vapor in the air. Above 60% humidity a variety of molds will grow, and dust mites thrive in high humidity environments. Low humidity – 30% or less – will make the air uncomfortable to breathe. If you can strike a balance – between 35% and 50% - you'll reduce the ability of molds to grow and you'll reduce nasal reactions to dry air. You may need either a de-humidifier or a humidifier to achieve this balance, and things can change between seasons.

Finally, back to ventilation. Besides having a filtered vacuum cleaner and not overusing cleaning chemicals, getting fresh air into your home is the first best thing you can do to reduce indoor air pollution. The cooler weather now makes this a delight. Open those windows and let that fresh mountain breeze wash through your home.

8

Appliance Tips and Tricks

Keeping Your Appliances Healthy Part 1

The appliances in our home are so taken for granted that we rarely think about taking care of them. The fact is, we can save money and energy by following some simple tips on daily use and maintenance. From refrigerators to water heaters, this multi-part series on appliance care will give you the information you need to reduce energy and expenses in your home by making your appliances last longer.

The appliances in our home - from the refrigerator and stove-top to the water heater - are so taken for granted that we are shocked when one breaks down. After dealing with the crisis, we tell all our friends about the problem the next day - "Can you believe it - the water heater just stopped working! Gee, it was only 22 years old! We had to take cold showers!"

When you think about it, the appliances in our homes and apartments hold up pretty well without any attention on our part. Most of these appliances, such as heating and cooling components, and kitchen appliances, chug along just fine without any attention. The fact is, all of these devices can last longer and save us money if we follow a few simple tips. Imagine that water heater or that compressor outside giving us a few more years of life and saving those repair dollars for when we really need them.

Take out a calendar and mark on it the few times a year that you'll go through this checklist. Spring and fall are always good times to do this, since we're taking care of other household chores and it won't seem as daunting.

In the kitchen: Pull out the refrigerator from the wall. Wow! Look at all that stuff back there! You'll find all sorts of things that became lost during the year, including Fluffy's mouse and Fido's chew toy. Vacuum and wipe down the coils, if your unit has them. You might see a vent cover - pull this off and vacuum it. Clean and vacuum the floor. This cleaning insures that air flow is good around the unit and the motor will last longer as a result. After you push the unit back in, check the door seals and wipe them clean. Now give the toys back to Fluffy and Fido. When they go missing, it's probably time to clean again.

Air conditioners and air quality appliances (such a de-humidifiers and air cleaners): As a home inspector, I find clogged air filters about 70% of the time. I am sure you are in the minority, aren't you, because I'm sure your filters are clean . . . but check them anyway. Make sure you find all the filters . . . sometimes there will be a return duct you forgot about. This is an item that is best put on the calendar monthly, because it's one thing that will really make a difference in your monthly electric bill. When the ducts have clean filters, everything in the system is more efficient, and the ductwork stays cleaner. If you have a through the window air conditioner, change this filter monthly also. You can buy disposable filters or filters that you can clean every month. You can also purchase a can of filter coating spray in most home improvement stores that will help the filter trap even more contaminants - and this spray combined with the cheapest filters will do a wonderful job keeping your indoor air clean and fresh.

Now go outside to where your air conditioning compressor is. This outdoor unit - the compressor - depends upon air exchange to work efficiently. Take a garden hose and clean this unit along the sides where you'll see what looks like a radiator. These are called coils and dirt buildup here should be cleaned off. Check around the base of your compressor unit and clean away any debris that you see - leaves, rocks, and anything else that is in the way of air flow.

To make sure your heating and cooling appliances serve you well year after year, consider a maintenance agreement (usually about $150 a year) with a good local HVAC company. Regular service checks (Fall and Spring) will be the other half of your own good care and your equipment will last longer and cost less to operate.

Keeping Your Appliances Healthy Part 2: Water Heaters

In last week's column, we talked about how we take our home appliances for granted because they rarely fail us. Why not get even more life from them? All of these devices can last longer and save us money if we follow a few simple tips.

Did you know that your water heater can last longer - often a lot longer - with some routine maintenance? I know that some of us "install it and forget it" and say, "we'll buy another one if it breaks" because we don't want to be troubled with maintenance chores. Stop and think about it. If you pay a handy-person to clean the sediment out of your tank every year and you get 5 extra years from your appliance - that's a real savings. The cost of new appliances has gone up, so consider the smaller maintenance cost to keep things running longer.

You have heard this before and its true: your water heater should be drained (several buckets worth of water after you empty it) regularly to minimize sediment in the tank. If you're handy, you can do this yourself. Find the booklet that came with the tank and follow the instructions. This sediment is a mineral deposit buildup which reduces the efficiency of the tank, costing you more money to heat the water. That cracking and bubbling noise you may hear when the unit is heating up is the sound of air bubbles between the sediment layers turning to steam and bursting through them. Sediment can drift into recirculation lines, jam open check valves, and cause the recirculation pump to stick until it burns out. Filters and water softeners will help, and of course your water source makes a difference; if you have a well, and the water is not filtered, you will have more sediment, sooner. Safety tip: Make sure you see a line that runs from the top of the tank over the side and either out through the wall or to within 6 inches of the floor. This is the "TPRV" line, or Temperature Pressure Relief Valve, and it is designed to carry scalding water safely away from any persons standing next to the water heater if the unit should overheat. As a home inspector, I see these missing all the time. If yours is missing, have a plumber install one for safety.

Another problem some people report on their water heaters is a rotten egg smell coming from the water. Hydrogen sulfide gas (H_2S) is a result of bacteria growth in your hot water heater. If the smell is coming only from your hot water tap - then that is exactly what is happening. The warm temperature, the sediment, and chemical reactions create this condition. Don't worry, these bacteria are not harmful to you, but you will want to get rid of it by having your tank serviced. Once again, follow the manual or hire a professional - they will turn the temperature up in the tank to kill the bacteria and/or they will treat the water with a disinfectant (chlorine).

The average lifespan of a water heater is 8-10 years with no maintenance, and 10-16 years with regular maintenance. Some brands are built tougher (cost more) and have longer life spans. I have inspected water heaters that were still functioning after 30 years! This is certainly the exception, but it is possible. If you don't know the age of your water heater, take a look at the label on the side of the tank. Most manufacturers will put the month and year into the first 4 digits of the serial number.

Water heater energy saving tip: install a timer in the electric line to the heater and only heat the water when you know you will need it - in the morning for showers and at night for dishes or showers. Set it so it turns on about 40 minutes before you need it. This will greatly extend the life of the heater AND will save you 30% or more on your water heating bill.

Keeping Your Appliances Healthy Part 3: Laundry Appliances

In Part 2 of the series we talked about how to keep water heaters operating longer and saving you money in maintenance costs. Now we'll talk about appliances in your laundry room along with some myth busting facts about getting your clothes clean.

But before we get to the laundry, one more tip on your water heater. Water heating is our home's highest energy expense after our furnace/air conditioner. If we follow energy saving tips here, it will make a substantial difference in our electric or gas bill. Do you know the temperature of your hot water? Consider a balance between that scalding hot shower and your wallet; the lower the temperature, the less money it takes to heat. While 120 - 125 degrees is considered a standard, you may be able to get by with it set at 115 degrees (there is an adjustment on the heater itself - check your manual). Where is your water heater located? If it is in your crawlspace, consider insulating it if it is not insulated already. Last week we talked about installing a timer. Why spend the money to have hot water available throughout the entire day if you're not using it? Finally, if your home has "instant" hot water which circulates constantly so you have that water available at all times, consider unplugging the circulation pump, or putting a timer on this pump. Instant hot water is a great convenience, but we do end up paying for this convenience.

Now to the laundry room. The water heater we just talked about plays a big part here too - 80% of the energy used to wash clothes is heating the water! If you are a Consumer Reports reader, you know that they have tested detergents in cold, warm, and hot water - and not found any difference in cleaning power among the three temperatures. I know this is counter-intuitive; we just "know" that more heat and more detergent have got to get those clothes cleaner, because we have that experience in the shower ourselves - hot and soapy gets US clean - but that's not how your clothes get clean. In fact, using too much detergent when washing your clothes will deposit an unpleasant residue and leave everything dirtier than when you started. So, follow the recommendations for your washer, and read the instructions on the box of detergent - modern high efficiency appliances can clean a load with about a quarter of a cup of high energy "HE" detergent. Use cold water and save some energy. Don't worry if you do not see any sudsing - "HE" detergent is designed not to create suds.

Don't overload your washer. It's tempting to just stuff whatever the laundry bin has accumulated over the week into the washer, but the clothes will not be able to expose all surfaces to the washing action. Stuffing the washer will reduce its life by stressing the drive mechanism and it can break early.

When your washer switches to the spin cycle do you feel as if someone is attacking your home's foundation with a jackhammer? This is a common complaint in our typical "stick-built" mountain homes where our washer and dryer are not on the first floor sitting on concrete. Two things you can do to reduce this effect are running the spin cycle on a lower setting - medium and not high - and putting foam feet under the corners of the machine. Sometimes machines are installed without leveling them - a big cause for shake - so make sure your machine is leveled and that the feet are secured tight against the adjustment nuts. Follow the manufacturer's recommendations.

If you have a front-loading washer you may have noticed a moldy smell coming from it between loads. This is common and calls for a little extra care. Because the front loader traps water in the seal around the door, you should take a little bleach and wipe the inside of the door seal to get rid of this old water. You can run a quarter cup of bleach through a cycle as well, however, some of our fancy septic systems do not do well with bleach. In this case, the wipe down should work fine. Finally, if you don't have small pets or children, you should try to leave the door open to the air to reduce potential mildew buildup in the washer.

Keeping Your Appliances Healthy Part 4: More About Laundry Appliances

In last week's column, we talked about how to keep our clothes washer humming and save money at the same time by reducing energy consumption. This week we'll talk about your clothes dryer and cover some important safety tips.

Before we do, lets cover one more item about your clothes washer. I received a call from a Progress reader who had a question about using cold water for the laundry. She raised a good point about our well water here in the middle of winter - COLD - really cold! Although saving money by not always using hot water is a good idea, you should use your judgment. If you have a family member in the home who is ill, or if you are washing bed linens, you should use hot water - preferably 130 degrees - or use a little bleach to disinfect the water to reduce the spread of winter germs and allergy dust.

While clothes washers have an average life expectancy of about 10 years, clothes dryers have a life expectancy of 12-14 years. I know we all have a story about a washer or dryer that just wouldn't quit and functioned well beyond these timeframes - "Aunt Nellie's Maytag" that went for 20 plus years without even a single service - but this is the exception. With a bit of care, we can maximize the life we get out of our clothes dryer.

We said it about washers, and we'll say it about dryers - don't overload the machine. In addition to stressing the components, the clothes will take longer to dry. Realize that the dryer you have, even if it's a matched set with your washer, is not necessarily designed to take the size of your wash load and dry it properly. You will alleviate this problem by not overloading your washer in the first place. A larger load in the dryer will take much longer to dry and use more energy than drying two smaller loads. Restricted air flow in the dryer will also reduce the lint (cat and dog hair?) that gets sifted out of your clothes as well as tax the motor and belt, reducing its life. Use a vacuum cleaner and attachment to remove as much lint and dust from the interior of your dryer as you can every time you vacuum around it - it won't seem like an extra chore this way.

Lint from your clothes is an enemy in more ways than you might think. If you forget to clean out the lint filter, energy efficiency is greatly reduced AND your clothes will not come out as fresh as they could otherwise. Worse, excess lint could become trapped in your dryer vent line and, with the right conditions of clogging and temperature - start a fire. The Consumer Products and Safety Commission ("CPSC") reports that there are over 15,000 fires every year caused by lint overheating in the dryer vent! To avoid this from happening, you should check your dryer vent. Is there clearance for it behind the dryer so that it's not crushed? What material is it made out of? You should be using only metal ducting, not plastic or thin foil. Where does the duct go? Hopefully straight out the wall, but sometimes our laundry room is not located on an outside wall, and the duct has to travel some distance to make it outside. If this is the case, you might need a vent booster to get the ventilation power you need for your dryer. Go look at the vent outlet outside your home. It is clear of debris or clogged? If it's even partially clogged, not only will your dryer wear out faster, but you'll use much more energy, and you'll risk having a fire start in the vent line. As a home inspector, you'd be amazed at what I find - from dryer vent outlets that are below grade (covered with mulch) to vents with no door or slats and assorted animals living in them, to no vent at all. To check your vent, start your clothes dryer and go outside to the outlet. You should see the blower door or the slats open and a good strong volume of air escaping. If you don't, consult your dryer manual for the best way to clean the vent, or you can hire a professional to solve this problem for you. Cleaner clothes and longer appliance life will reward you for your extra effort.

IMPORTANT SAFETY NOTE: In no case should your dryer vent terminate in the attic or into the crawlspace. If it does, you'll be less inclined to keep it lint free. If it does clog and overheat, your entire home could go up in flames. If your vent does not terminate outside, have a qualified person re-route it to the outdoors.

Keeping Your Appliances Healthy Part 5: In the Kitchen

This week we will begin talking about taking care of your kitchen appliances and cover some nifty tips and tricks to keep things clean in your kitchen.

There are two different types of dishwashers you will find in your kitchen. I will touch on the first one, and then spend more time on the second type. The first one requires completely different care from the second one. If you have the first type, you may need to spend more money taking care of it. This care will probably come in the form of compliments, appreciation, and dinners out at restaurants. What in the world are we talking about? You guessed it - this type of dishwasher is human, and requires coaxing and kindness to keep it going. This dishwasher might even be yourself - if so, take yourself out to dinner! The second type of dishwasher is the mechanical, sometimes called "automatic" dishwasher - and with a little care will give you long and faithful service also.

Run your dishwasher at least once a week to keep it clean and the hoses flexible. Periodically look in the bottom where the drain is to make sure there's nothing resting there (like silverware) or debris caught there. Check the spinning arms in the middle to make sure no holes are clogged. If they are, you can use a piece of wire or needle nose pliers to clean it out. Then put one cup of distilled white vinegar in a coffee cup in the top rack (upright) and run the dishwasher without anything else in it through a full cycle. You'll think your dishwasher is brand new! Wipe down the sides of the doors, where all the coffee ends up dripping.

How hot is your water? The dishwasher needs at least 120 degrees to get your dishes clean; 125 degrees is even better. Some units will heat the water to 130 degrees or more on their own, which is a nice feature if you don't want your water that hot.

Are your glasses looking cloudy? Soak one of them in vinegar for 5 minutes. Does it look better? If so, your water is hard, which means it has a lot of calcium and magnesium ions that limit the ability of detergent to do its job. You might need a water softener to get the water back in balance. If the cloudiness does not go away, then the glasses are becoming "etched" - a permanent condition. In this case, you are probably using too much detergent, or pre-cleaning the dishes thinking it's a good idea - but a little soil is actually good for a dishwasher's cleaning action. Just scrape off the bulk of the food and put the dishes in the washer. Most modern dishwashers have a macerator blade in the drain that will chop up any food debris. As for detergents, a good dishwashing detergents regularly. If you have a septic system - most of us here in the mountains do - then you will want to use a low or no phosphate detergent in your dishwasher.

Lastly, take a look at that instruction manual that came with your dishwasher that you threw in the back of the drawer. It actually has some great advice to keep your unit running for a long time!

Back to that first type of dishwasher - you or your friend/spouse - here is a great tip following the emptying and cleaning of your kitchen sink - take a clean dish towel and wipe the sink out until it's shiny. Every part - from the faucet to the drain rim to the sides - get it completely dry. This will take you all of 10 seconds. If you do this religiously every single time you use the sink - you will never have water spots and your kitchen sink will always sparkle. Your sink will be the envy of your friends and neighbors and they won't know how you did it.

Keeping Your Appliances Healthy Part 6: Small Appliances

Last week we talked about dishwashers; this week let's talk about some of your small appliances in the kitchen and laundry as well as a few tips and tricks on cleaning.

Do you rely on coffee in the morning to get you going? I sure do, and if the coffee maker broke down I'd probably go look for a 24-hour convenience store so I could have my 6am charge. We tend to take our appliances for granted until they break. However, a coffee maker just declines gradually until we wonder why it takes so long to brew, or the coffee tastes bitter. You guessed it - it's time to clean it! What happens is that the sediment and scale build up in the lines, vents, and tank. To get rid of this build-up, run the coffee maker through a cycle with a vinegar solution comprised of one part white vinegar and two parts water. Now turn off the unit and let it sit for 30 minutes. Next, empty the solution down the drain or put it in a spray bottle and use it as a cleaner in the first round of some of your dirtier jobs (vinegar is a terrific cleaner, and gentle on the environment). Now run two cycles of water through the machine. Wipe the outside with the vinegar water solution and dry. You'll think you have a brand-new coffee maker. Some sources say to do this every week - but unless you're running a boarding house, every few months should do it.

What else can you clean with vinegar? The possibilities are endless, given vinegar's acidic and naturally antibacterial cleaning power and its relative chemical safety. In fact, those of you who know how to keep your air conditioner's condensate line unclogged with bleach - switch to vinegar and it works just as well, and is much more environmentally friendly. All of the HVAC (heating and cooling) techs I have spoken to have now gone to vinegar, saying it is not as rough on the equipment.

Is your steam iron slowing down? Fill the water chamber with equal parts of white distilled vinegar and water. Set it upright on a safe surface and let it steam for 5 minutes. Let the iron cool. Now refill the iron and shake it in an old towel. Next time you fire it up, make sure you iron an old cloth for a few minutes to exit the rest of any remaining vinegar before moving on to your clothes.

To see hundreds of other uses for vinegar, just type "cleaning with vinegar" into your favorite internet search engine, and see the endless advice that comes up.

Luckily, most of our kitchen appliances keep going with no maintenance or help on our part. So, keeping them clean is probably the most we need to do, along with following safety tips. One of the things most us resist is reading the manual that came with the device. "No time!" we exclaim, but then wonder why the device is not doing what we want it to do. Then we actually waste more time looking for the manual than it would take to have read it in the first place. So, do yourself a favor, grit your teeth and READ the first few pages of that manual, and then put it in a drawer or box with ALL the other manuals. Admit it - I will - all the manuals are NOT in one drawer, they are scattered all over the house - in the strangest of places, like the bills drawer or the magazine drawer, or the tools drawer; the garage, the attic, even in the car. If we all start now, we might be organized by Christmas 2035.

Take the Appliance Facts Quiz

Several readers wrote to say they enjoyed the quiz last week. Here is another quiz, this time on interesting appliance facts you may not know. Play "handyperson" and see how you do.

1. The crackling and bubbling noises that my water heater makes is nothing to worry about. True or False?
2. What do I do about the bad smell that emanates from my garbage disposal? A. Ignore it, its normal. B. Throw in some potato peels. C. Grind a lemon wedge and four or five ice cubes. D. Use an old dish washing brush and brush the inside of the disposal cavity.
3. You smell rotten eggs coming from your hot water tap. What could this be? A. It's not coming from the hot water tap. It's really the garbage can. B. The septic system is backing up. C. It is green mold growing in your hot water line. D. It is bacteria growing in your water heater.
4. The biggest problem that inspectors find with heating and air conditioning systems in homes is: A. Not enough vents are open. B. Thermostats are set too high. C. Mold is growing in the ductwork. D. Filters are clogged.
5. The biggest mistake we routinely make when doing the laundry is: A. Money laundering. B. Filling the washer with too many clothes. C. Using too much detergent. D. Forgetting to take stuff out of the dryer.

ANSWERS:

1. False. That cracking and bubbling noise you hear when the unit is heating up is the sound of air bubbles between sediment layers turning to steam and bursting through. Sediment can drift into recirculation lines, jam open check valves, and

cause the recirculation pump to stick until it burns out. The solution? Periodically drain the tank to remove the sediment. Your water heater will last longer.

2. C. and D. Ice sharpens the blades; lemon or lime freshens the blades. Using an old brush with some dish detergent (keep separate from the regular brush!) on the inside cavity will clean the scum off and it will stay fresher longer.
3. D. Hydrogen sulfide gas (H_2S) that smells like rotten eggs is a result of bacteria growth in your hot water heater. If the smell is coming only from your hot water tap - then that is exactly what is happening. The warm temperature, sediment, and chemical reactions create this condition. Don't worry, these bacteria are not harmful to you, but you will want to get rid of it by having your tank serviced. Hire a professional - they will turn the temperature up in the tank to kill the bacteria and/or they will treat the water with a disinfectant (chlorine). In the future, make sure your water heater is set to a minimum of 120 Degrees °F. I set mine at 122 °F to be on the safe side.
4. D. Filters are clogged. As an inspector, I find this condition 70% of the time. If you want to save money and reduce air pollution in your home, simply changing the filters every few months will do wonders.
5. All answers are correct. We routinely forget to take things out of pockets; we overfill the tub and get less than satisfactory results; 90% of us use twice as much detergent as the load requires, and we've all discovered a wrinkled mess in the dryer. Read detergent boxes carefully and sort loads into smaller batches. Modern washers use a fraction of the water and energy that machines used to use so don't skimp on doing more washes with fewer items.

Appliance Tips You Already Know But Avoid Doing 1

There isn't enough time in the day to do everything you want. I'm not sure how this happened, but for me it was about 10 years ago. In a flash, I realized that time was not the reliable scientific constant I thought it was. Suddenly the day consisted of 18 hours instead of 24, or so it seemed. And part of that time is for non-productive sleeping. Things have only gotten worse since that stark realization. Either the time is really going faster, or there are more things to do. Somehow, I think it's both.

Are you in the same predicament? If so, it will save time if we perform some of these basic "keep it going" tasks with our appliances. Nowadays service calls are expensive. While the average lifespan for compressors, refrigerators, and washers is more than 10 years, let's go for the "more" by putting the following checklist items on our weekend To Do List.

Clothes dryer lint. If your clothes dryer does not have an internal lint filter that is obvious, put a large text reminder inside the door where you will see it every time you put a load of laundry in the machine. Clogged internal filters will greatly reduce the drying power of your machine because the air flow will be reduced. If an internal filter is excessively clogged, a fire could actually begin in the machine. According to the National Fire Protection Agency, over 15,000 fires occur in clothes dryers every year. Also check the exterior vent every few loads. The "out of sight out of mind" nature of this outdoor vent can create lint clogs that also trigger fires and shorten the life of the appliance. When checking the outside vent, make sure the flapper door is working so you don't encourage birds to adopt your vent as their home.

Refrigerator door seals. Place a dollar bill between the door seals and close the door. If you can pull it out easily, your seals are shot or they have lost their magnetic properties. This causes the compressor to work overtime trying to cool the interior and shortening the life of the motor, and will burn up your electricity dollars. If you're handy you can install a new seal yourself or clean and re-magnetize the old one. If the refrigerator is over 12-15 years old, you might want to consider a new model, which will be more energy efficient. Prices are reasonable compared to repair costs, which have skyrocketed.

Air conditioning filters. If you have ever rented a home, you probably got a maintenance checklist from the landlord. The top item on this list is always air filters. In fact, one landlord inspected our rental monthly, and specifically the furnace/air conditioning filters. This is because these filters are the "lungs" of your home and there is a direct relationship between your health and how clean these filters are. Heating and air conditioning filters trap all those nasty things you know are in the air and on the floors, including pet hair, dust bunnies, pollen, carpet and furniture chemicals, and bugs.

The second reason to check heating and air conditioning filters every month and replace at least every 3 months is to maintain the health of your heating and air conditioning equipment. Blower motors can go trouble free for more than 15 years if you keep things clean.

Appliance Tips You Already Know But Avoid Doing 2

In the last column, I talked about some of the simple but important things you can do to save time and money with your appliances. We have enough things to do once the weekend rolls around than to spend time repairing or buying washers, dryers, refrigerators, and air conditioners. Here are some more items to put on your to do list to keep things running year after year.

Front loading clothes washer seal. If you have a front-loading washer, you already know that the seal around the door can trap water. To keep this area from getting moldy and smelling like a rotten tomato went in with the laundry, spray some white vinegar and water on a cloth and run it around the seal after the wash. A light solution of bleach and water works well too. If you can leave the door open between washes, all the better. If you have young children in the house, it is a better plan to wipe the seal and close the door.

Dishwasher filter. For years, I owned dishwashers and had no idea that there was a filter in the bottom of the machine. When I finally read about it in a course I went to on home maintenance, I was surprised. I thought to myself, why have a filter in your dishwasher if the machine is getting everything clean to begin with – wouldn't the filter be clean too? It turned out to be an erroneous notion, since the filter's purpose is to trap big weird food things like broccoli stalks that your plumbing can't handle. Some modern dishwashers also have a macerator – a food chopper – that acts as an internal food disposal. Even these units have filters. The good news is that these are very easy to check and clean. Simply remove the lower rack off the track and twist or lift the filter out and take a look at it. If it is gunky, just rinse it under hot water and put it back. Doing this will keep water flowing and the dishwasher will smell better.

Garbage Disposal. Does your disposal constantly smell bad? Just throwing a lemon and some ice cubes in there doesn't work for long, and here's why. The seal at the top that is there to keep silverware and your hands out of the disposal trap moldy gunk on its underside. Take an old dishwashing brush (keep in a special place so that it is only used for this job) and place dishwashing soap and a little bleach spray on it. Drop it down into the cavity and brush the sides under the seal, up to the seal, and then down onto the blades. Do NOT run the disposal while you are doing this. Remove the brush. It will be a mess. Rinse thoroughly with hot water and then put back in your special spot.

Now you can run the lemon and the ice cubes with the disposal running for a minute. This trick should cause the sink area to smell great for another week or so.

Appliance Tips You Already Know But Avoid Doing 3

In the last several columns I discussed simple tips to keep your household appliances trouble free for as long as possible. Here are some final tips on appliance care, and how to save money on repairs.

Heating and air conditioning ductwork. I have had people write and ask if they should get their ducts cleaned. The answer is no – with some exceptions. Exceptions include unusual circumstances, such as water getting in to the ducts, vermin getting in to the ducts and becoming petrified (that's fun to find), or humidity problems causing mold to be visible in the vents. These special cases warrant a professional inspection and cleaning. Otherwise, don't worry about it, but do consider the following advice.

Vacuum the area inside the cavity where the air filter goes, and vacuum inside the floor vents. These vents are not fastened down for this reason. You might find some interesting things in there. As a home inspector I would take these floor vents off to look inside the ductwork. I can tell a lot about what the maintenance has been on a house by looking inside these registers. I have found everything from toy trucks to jewelry (returned to the owner) to cash (also returned) to petrified mice and lizards (not returned).

Check washing machine hoses. Most of the water supply hoses for washing machines are reinforced rubber, and the quality level varies widely. We forget that they are there until one bursts. If this happens when we are not at home, the force and volume of the water will do tremendous damage. As an inspector, I would always check these hoses, and would find them highly deteriorated about a fourth of the time. Your maintenance checklist should definitely include replacing these – an easy job – every 4 to 5 years. Don't skimp on quality – consider stainless steel braid reinforced hoses. As a precaution, turn the water supply off if you are going to be away for more than a few days. Not that a lot of damage can't be done when you're at work each day – but it does seem as if Murphy's Law (what can go wrong will go wrong) will activate as soon as you leave on vacation. Some washing machine manuals actually tell you to close the water valves after every wash. Most of us won't do this, but it's not a bad idea.

Saving money on repairs. Combinations of electricity, electronics, and plumbing can make modern appliances difficult to repair by ourselves. Unless you are an accomplished handyperson with access to instruction manuals, hire a professional to repair your equipment. When you get quotes, find out what the minimum charges are, and if it includes travel time. Consider that a factory trained service tech with a fully stocked truck at $75 an hour is going to be a better deal than a handyperson at $25 an hour if they have to first diagnose, then go buy parts, and then make repairs. Work quality, inventory, and experience are what you are looking for. Ask your friends and neighbors who they use.

Probably the biggest tip I can offer is for you to actually read the owner manuals that come with your appliances. Make your checklist, and then place these manuals in a file where you can find them easily.

Dishwasher Tricks: 5 Things You Might Not Know

Dishwashers wash dishes, right? Yes, and more. If you own a dishwasher, here are 5 tricks you can use to save time cleaning things other than dishes.

First, take a good look inside your dishwasher. When was the last time you looked? Find the debris filter on the bottom of the tub. Is it clean? Make sure there are no toys, utensils, or food pieces in or on this filter. Most filters pop out for inspection and cleaning. Next, take a cloth or sponge with a little bleach on it and wipe down the seals along the sides of the dishwasher. This is a breeding ground for mold and bacteria.

Now that your dishwasher is clean, try putting the following things into the dishwasher and running them on a regular cycle.

Glass lamp and light globes, vases, ceramics, and other glass or ceramic artwork or fixtures you don't feel like cleaning by hand. Place these in the washer by themselves with plenty of room between items and run through a regular cycle. The light fixtures will look especially great when you're done.

Plastic hairbrushes (not with natural bristles or wood handles), combs, and other plastic do-dads – place in a mesh bag in the upper rack, where it's not as hot. Until I discovered this trick I was soaking these items in soapy water with a little ammonia . . . what a pain. Other plastic items include the bacteria laden dog and cat toys, and plastic children's toys.

Ever wonder what is lurking on your toothbrush? Throw that in too. Look around the bathroom – you will find all kinds of things that need cleaning. Use common sense though; obviously, you do not want to put anything into the dishwater that has any internal electronics or an external power plug.

Kitchen sponges, brushes, and appliance parts. Sponges trap bacteria and mold, as does the dish washing brush – the dishwasher will sanitize these. You can run these in the regular cycle on the top rack. Also load up grates, range burner (gas) caps, microwave rotation plates, aluminum mesh filters from the microwave or range hood, and the plastic bins and holders from your refrigerator.

Clothes. Clothes? Ok, not really. But how about cotton hats – baseball caps, sun caps, and other similar items – do these separate from the dishes, using borax in the dispenser cup, and skip the high heat wash and the heated dry cycle. Put these items on the top rack, and when complete, reshape and let air dry. I have been thinking about loading all of my family's sneakers into the dishwasher to see how they come out, but I have not tried this yet!

Your dishwasher uses a lot less water than your washing machine. Modern dishwashers use 4-8 gallons of water in a cycle and your clothes washer uses 15 (front loader) to 45 (older top loader) gallons. Don't worry about running it to handle these tasks. Just think it through to make sure the item won't melt, destroy electronics, or get caught in the spinners. Be creative and let me know if you come up with any new ideas!

More Dishwasher Tricks

Do you own a dishwasher? Not the human kind, but the appliance kind? Here are some handy tricks you can use to wash unusual items in the dishwasher. If you have only the human kind – well, you can try to give them these jobs too, but be prepared for some grumbling noises.

Non-human dishwashers use a lot less water than your washing machine. Modern dishwashers use 4-8 gallons of water in a cycle and your clothes washer uses 15 (front loader) to 45 (older top loader) gallons. Don't worry about running the dishwasher to handle extra tasks. Here are some interesting ideas to try.

- Air conditioning filters, mesh screens, and vents. First vacuum off the debris. Then run in the top rack – small pieces should be in a mesh bag. Run the rinse cycle with no heat.
- Dustpan and other plastic cleaning items. Clean off loose debris first.
- Desk accessories. This includes pen and pencil holders, lamp globes, and mouse pads.
- Fake flowers. Have you taken a good look at the plastic flower arrangement in the corner of your dining room? If you've tried cleaning the dust off the leaves, you've discovered that they are sticky. Cleaning each leaf by hand will take a long time – instead, place them into a mesh bag and put in the dishwasher for a regular cycle without drying heat.
- Oven knobs and other ceramic do-dads. These clean up nicely in a top rack rinse.

- Car parts. What's grubby? How about cup holders (usually removable), coin holders, small hardware pieces (use a mesh bag), hubcaps, metal panels, etc. Avoid putting carpet in the dishwasher; it will probably disintegrate. No reason to wreck the dishwasher unless you want an excuse to buy a new one.

If you do some reading on the internet you will find some other strange things that writers suggest you try in your dishwasher that I am going to label "stupid tricks." They are stupid tricks because they are more trouble than they are worth, smelly, or could be damaging. Here are a few stupid tricks:

- Cook salmon. Are they kidding? Have you noticed the last time you cooked salmon and just put the dirty dishes into your dishwasher, the appliance smelled fishy?
- Computer keyboard. What? Put delicate electronics into the dishwasher? Didn't someone say it was a bad idea to spill liquid on the computer keyboard? For about $15 you can have a brand-new keyboard that doesn't have the letters worn off delivered overnight from Amazon.com.
- Switchplates. By the time you search out your toolbox and discover that the small flat headed screwdriver is missing, and find it, and then go back to the switch plate, drop one of the tiny screws, get down on your hands and knees with a magnifying glass to find it, then put the plate in the dishwasher, you will have spent 20 minutes doing something that a rag wet with dishwashing liquid would accomplish in about 45 seconds.

If you think of other tricks or have tried unconventional cleaning tasks with your (non-human) dishwasher, write and let me know for future columns.

9

Organizing and Cleaning Tricks

Secrets of the Well Organized

A few Progress readers wrote to me last week saying that they are inexorably messy people, and that even the exotic organizing tricks I wrote about are beyond their ability. They went on to say that they can find anything anywhere – because their things are everywhere. They said they feel guilty about being naturally messy.

There is definitely a case to be made for messiness. Why not? Why care about what things look like? The answer lies in how we were raised and in our natural personalities. If we were told constantly to pick up our toys and put them away neatly, then we are going to be either messy to be rebellious, or neat to continue the habit. Some personalities tend to be more detail oriented and care more about what things look like.

If you are the maverick amongst us who likes to throw things down and hope for the best, don't let me dissuade you from your habits. Unless there is a good reason for you to organize, and you can find what you need when you need it, don't suffer a guilt trip. Of course, other people in the house may tease and nag you, but stand your ground.

The neat-nicks amongst us will argue that things are not only easier to find when we are organized but it provides for a better environmental plan with fewer unnecessary items around the house and a better mental outlook knowing that the things you don't need are going to people in need.

For the neat-nicks, here are the keys to being organized.

- Mental approach or attitude. It's not what organizers you buy, it's being disciplined about only keeping what you use and love. In a best seller book on organizing from Japan, Marie Kondo says that we should only keep something if it "sparks joy."
- Set aside some quiet time to stand in each room of your home with a notebook. Let the scene soak in. Is everything you see contributing to the overall picture of beauty and function? What is extra, what is missing, what is out of sync? Write down what you see. When you have completed this exercise, you may find yourself calling the charity truck for a pickup. Don't do more than one or two rooms at a time, or you'll feel overwhelmed.
- When you feel that your world is organized and where you want it, keep it that way. To do this, clean as you go and put things where they belong immediately after using them. Rather than putting lots of little things in organizers or keeping them out in view, put them away in drawers, cabinets, or trunks. Yes, you CAN teach children to do this.
- Make constant decisions about what delights you, and only keep those things. If this scares you, then put things in a closet for a few weeks to experience what it's like without it. You can always pull it back out if you really love it.

For more organizing philosophy and tricks, pick up a copy of "The Life Changing Magic of Tidying Up: The Japanese Art of Decluttering and Organizing," by Marie Kondo. If the messy people have read this far, this is a signal that there is hope for you.

Junk Drawer Heaven

Do you have a junk drawer? If you don't, then you are either a two-year-old or you are blessed with the no-pack-rat gene.

"Honey, where do I find a screw to fit this bracket?" calls my husband from the kitchen.

"In the junk drawer!" I yell from upstairs.

If you are a pack rat, then it's more likely you have a "junk closet" or even a "junk room" (called a garage or pantry). Do you sometimes feel that you have become disorganized or have difficulty finding things? If you're just starting out in a new place, then you don't have to worry much about this problem, unless you are moving from an old place. If you are moving from an old place, then you're likely to replicate the same junk problem in your new place. Whether you have junk or are about to start creating junk, here are some tips to become more organized.

First, I have to tell you that I am a junk drawer advocate. If you are over 18, you need junk drawers. If you are excessively organized and don't have junk drawers, then you don't need to read any farther. You are, in scientific circles, what they call an "outlier", i.e., not normal.

Junk drawers happen. They have a mind of their own. They are a convenient area to hide things before guests come over, for cleaning things off counters, and for putting miscellaneous things in that you don't know what to do with. Eventually . . . the drawer won't close.

So, secret number one is to have more than one junk drawer. Ideally, you should have a junk drawer in every room. This obeys the rule of "keep it where you use it."

As you use the junk drawer, watch for an accumulation of like things, such as screws, rubber bands, bag clips, and hardware. If you have another drawer with the same stuff, consider putting "like with like" unless you need the duplicate items in another area.

Secret number two is to place some baggies and masking tape with a pen in each drawer. When you find yourself throwing something into the drawer, like a charging cord, put it in the baggy and write down what it goes to. I wish I could time travel back to 1985 and do this for all the chargers I have. This technique also works well for toy parts and miscellaneous electronics and hardware.

Secret number three is to sort the drawers twice a year. With small boxes and baggies handy, remove the things you are not using and put them together from all the drawers. If you're the pack rat ("Honey, I know I am going to need this broken blender base someday"), then all of these items will go to a marked box in the garage or attic. If you're not the pack rat, then see what can be recycled, thrown out, or donated to the thrift store.

Lastly, purchase several clear plastic drawer sets from your local household goods store and place these in strategic locations for tools, screws, hardware and glues, and other categories such as batteries and electronics. The possibilities are endless. If you are lucky enough to have a few closets, you can designate an entire shelf to these special organizers.

Looking for something? Now you'll know just where to look.

Checklist Power

Flying an aircraft is one of the safest things you can do. Being up in the air and having something go wrong is life threatening. It's not like being in your car, where you simply pull over to the side of the road. So why is flying an airplane a lot safer than driving a car?

Checklists. Pilots use checklists for everything. Pilots use checklists for training, for preflight, for takeoff, for landing, and post flight. There's no doubt that if we all used checklists for driving, it would be a lot safer.

I am a fan of checklists (I'm also a pilot). Over the last few weeks in this column I've talked over and over again about having checklists for maintenance in your home. Spring checklists, fall checklists, appliance checklists, safety checklists; there are dozens of checklists.

Checklists drive efficiency, safety, and accomplishment, but only if they are used. It's easy to skip things or just leave the list on the shelf. "I'll remember," you say to yourself. This even happens to pilots. One day I was at a busy municipal airport getting ready to fly when I heard a pilot, who was inbound for landing, describe a problem with his landing gear. It was jammed and would not drop down and lock. Someone on the ground took a look as the pilot flew overhead in a fly-by. He radioed the pilot, saying, "Your wheel chock is jammed between your nosewheel and your gear door."

Fortunately, the story ended well with a belly landing that everyone walked away from. But it was a big embarrassment to the pilot, who obviously had not followed his pre-flight checklist and removed the chocks and chains from the wheels.

In our homes, not using or following a checklist may not render such serious consequences, but using one will help us avoid problems in the future. My recommendation is to have several loose-leaf notebooks. Label them, "Appliances," "Safety," "Seasonal," and "Organizing and Cleaning." Then, use the checklist items I've been writing about, or type "Household Checklists" in to the internet. You will find an abundance of ready to use lists to customize and place in your binders.

As humans, we know that it is relatively easy to forget something or make a mistake. Checklists are powerful for more than just flying or draining the water heater to get the junk out of the bottom. The other day my husband forgot to leave work for an appointment. After realizing the error, he was upset and commented, "But, I wrote it down!"

"You wrote it down but how do you know to look at where you wrote it," I said.

I think I made him more upset. "Ok, he said, "I will have to figure out a system."

And that is exactly how to use checklists. Use a system.

- Answer how you will use the list (example, loose-leaf notebooks with lists)
- Answer where you will use the lists (have one spot for them)
- Answer when you will look at your lists (have a calendar for checks)

Using a checklist system in a dedicated way will save you time, money, and aggravation. Just think of yourself as a pilot.

Spring Cleaning Fun

I named this column "Spring Cleaning Fun" because I wanted you to begin reading it. If you enjoy spring cleaning, then you are in the minority. The only "fun" in spring cleaning is the possibility of getting outdoors when the sun is warm and bright and the smell of new foliage drifts through on the breeze as you beat the dust out of your rugs.

Now that I have your attention, let's figure out how to make spring cleaning fun. You have some options. You can simply forget it entirely and watch a rerun of "I Love Lucy," or you can get organized and ease through the process in the least amount of time. Here are some suggestions for the latter.

Pull out your checklist and do a little pre-planning. If you don't have a checklist, go to Google and type in "spring cleaning checklist." About 10 good lists, all with about the same chores, will pop up. The whole idea behind spring cleaning is to hit the areas that you have not even thought about all year. These include:

- Kitchen: Vacuum and clean behind the refrigerator (clean the refrigerant coils too if your refrigerator has them), clean and organize cabinets and drawers, throw out expired products, clean the oven and hard to reach spots on woodwork, baseboards, under tables.
- Bath: Throw expired stuff out of the medicine cabinet, clean the fan and vent, wash the shower curtain (or wipe down if it's plastic). Clean in nooks and crannies, dust light bulbs and fixtures.
- Bedroom: Flip over your mattress, wash blankets, pillows and covers (follow directions), clean drapes and blinds, clean under beds.
- Closets: Inventory your clothes. Be ruthless and give away the things you have not worn in years, unless they hold sentimental value. Organize what remains by season. Dust shelves and clean woodwork.
- General: Clean woodwork, baseboards, hard to reach places, and fans and vents.

That's just a starting point. Look over the checklist you printed out from the internet and delete or add items. Here's how to get started and stay motivated.

- Declutter the rooms first. This will save you a lot of time as you clean. Put items away, throw away what you can, and put the yard sale stuff in the attic or garage.
- Get help. Sign up the kids, the neighbors, the spouse, and/or your friend(s) to come over and chip in. Promise them some

reward. You will have to deliver it, or you'll be doing all these chores by yourself next spring.

- Put on your favorite music.
- Assemble the cleaning materials.
- Wear fun old clothes.
- Take breaks to snack, have coffee, or sip your favorite beverage. Then jump back in.
- Pick the right day. Pick a sunny spring day when you can open the windows and really air everything out as you work.

Most important last item: when you are finished and everything is shining and organized, pat yourself on the back and congratulate your helpers! Go out to dinner or celebrate in some way. Rewarding yourself for tackling these tough extra jobs will make you more inclined to perform the same miracles next year. Have fun!

Bad Spring Cleaning Ideas

Why is spring the time that everyone talks about cleaning and organizing? This doesn't make sense. Spring is not the time to be indoors organizing and cleaning. Spring is the time to forget about your closets and your pantry and your drawers and get outside. Spring is the time to watch green stalks peek through the mulch after a misting rain. Spring is the time to rejoice in the renewal of the garden and the flowers. Spring is the time to feel the warm breeze on your skin and hear birdsong.

Over the last six weeks I have read many spring cleaning articles. Besides being the wrong season to bring the cleaning subject up – there are some authors who have gone out on a limb thinking up "smart" spring cleaning ideas.

But some of them are really dumb. Please, if you like these ideas, don't get mad at me. Go right ahead and use them. It's a happy world when we each let the other do clearly silly and counterproductive organizing tricks. Here are what I consider the best from the stupid list.

- Buy a full-length shoe hanger that covers the entire backside of a door and fill each shoe space with other stuff. This is a bad idea for 2 reasons. The first is that the hardware is visible on the front side of the door. These ugly hangers cry out, "There is a mess on the other side of this door that I didn't know what to do with." The second reason is that you will have so much "stuff" in these little fabric spaces you will not be able to find anything in there ever again.
- Paint aluminum buckets different colors and label them with the names of what is in them. Place these at the entry just inside of the door. You've got to be kidding. This idea clearly was for people with a "mud room" and two doors. If I lined up six or seven fluorescent buckets at our front door not only would I not be able to get into the house, but my husband would call the people in white coats to come get me.
- Install chalk board over the kitchen cabinets and hang a piece of chalk on each one. This way everyone can write down what they need for the grocery run, their "to do" list, what they want for Christmas, and messages for the Easter Bunny. Every family member can have their own writing space at their own level. Why not also have a space where guests can write their complaints or their praise? Lovely.
- Magnetic bathroom storage. You know the magnetic holder that goes on the wall in the kitchen to store large knives? Use this same idea to mount magnets in your bathroom next to your sink. Here are the many things you can load on to this rack: bobby pins, hair pins, tweezers, toenail clippers, scissors, and razor blades. This sounds very ugly and somewhat dangerous to me.

Here's one last idea that isn't so bad. In fact, I just may try it. Ice trays for jewelry! I can think of a lot of other cool stuff that would fit in these handy trays. How about screws, nuts, and bolts? Now we're getting somewhere.

Poor Organizing Ideas

In the last column, I talked about "spring cleaning" organizing ideas that appeared to be made up by writers who didn't understand the world that you and I actually live in. From fluorescent buckets lining the wall by the front door to fabric shoe racks making little things even harder to find – I am not sure how the authors came up with these ideas, but I was not impressed.

I found a few more of these counterproductive ideas. And, as I said last week, if you like these ideas, don't get mad at me. Go right ahead and use them. Thankfully we are in a world where some things will work well for some people and not for others.

In my mind the purpose of an organizer is to HIDE what you have while still knowing where it is. The purpose is not to ADD even more stuff to what you already have. It seems that the following ideas might just make things tougher to find and accelerate the rate that you accumulate things. Not to mention making your home look like a Friday's restaurant.

- A bungee storage wall. Inside your front door or in your pantry, throw up some old planks on the wall. Then go find all of the bungee cords you know are scattered about the house. No bungees? No problem, head over to Walmart. When you return, stretch the bungee cords from plank to plank using nails or hooks. Hang hats, newspapers, sneakers, clothing, and anything else that will drape over a bungee cord. Lovely.

- The kitchen is always a great place to make things look messier and busier. To achieve this look, go back to Walmart and find a large wall mounted pan lid rack. Hang this somewhere in your kitchen and fill it with all of the pot and pan lids that are taking up precious space in the lower drawers of your range or cabinets. Actually, I hope my sister does not read this because this is exactly what she has done in her kitchen. You can probably guess what I think about it.
- A similar idea for other areas – such as the bedroom, family room, guest room, or study – is a postcard rack organizer. I'm not sure where you will find one of these, I don't think Walmart will have it. A drugstore going out of business sale would be the place. Once you get your rack, brainstorm all the neat things that you can pull out of your crowded drawers and load into these postcard sized spaces. Line the bottoms of each space with duct tape, and then load photographs, pens and pencils, eyeglasses, keys, notepaper, grocery lists, books, scissors, and even postcards into the 24 or 36 spaces on this nifty rotating mess.

Ok, that's enough craziness. Next week I'll tell you about some smart organizing ideas that will reduce the mess, not add to it. Seriously!

Feeling Cluttered After the Holiday?

Are you feeling cluttered a few days after Christmas? Is there a lot of "stuff" lying around the house? Are you dreading putting all the decorations away? Do you want to exchange your family room for one in an empty model home?

Just about the time that your home is the messiest from all of the gift giving, all of the decorations, and all of the meals, you are probably not up to figuring out what your new year resolutions are going to be. I have a solution for both of these problems. De-clutter! Make one of your New Year's resolutions now and go ahead and do it before the new year even starts.

I hear you saying, "The end of December is not the time to start the spring cleaning." Why not? Think about it. When are you going to have time to do the spring cleaning? When might you have a few days off? Now! Not only will your decorations be put away neatly, but your home will look brighter and fresher, helping to stave off the winter blahs.

First, organize the gifts and make decisions about them before you put them in a drawer or closet. Will you use it? Really? If you know you won't, find a good place for giveaway and garage sale items and put it there. There is more "re-gifting" nowadays; if you are comfortable doing this, check the box for hidden "Love from Aunt Susie" notes and price tags. How embarrassing for you and your re-gifted recipient should you forget this step.

Next, look around the rooms in your home and ask this question: "Does this room look cluttered?" If the answer is no, move to the next room. If the answer is yes, look at the clutter. Is it too much furniture? Can you take some of it and store it? Is there anything you don't particularly like? Now is the time to give these things to the church or sell them.

Now, look at the "stuff" lying around. Books, newspapers, magazines, knickknacks, art pieces, etc. Is there too much? If so, perform two steps. First, ask yourself if you still want all this material in the room. If not, give it away or store someplace for your garage sale in the spring. If you really want to keep it, but don't want it out, imagine a new bookcase with doors or a storage chest that would go in the room to hide everything.

Removing objects from a room and placing small items in drawers, cabinets, or chests will make the room appear both larger and brighter. The storage technique makes it possible to be a pack rat without appearing to be one. The psychological effect of opening up room space like this cannot be underestimated.

This version of de-cluttering does not include closets because we cannot see into the closet mess. This assumes that you are able to close the door to the closet! If you cannot close the door to the closet, then you are going to need to make decisions about what is in there. Remember that it is the season to give, give, give.

If you can't stand the thought of doing this now, then cut this out and place it in a drawer labeled "Open in spring."

How to Tackle Spring Maintenance Chores

Afternoons are a little warmer and nights are pleasantly cool. Tiny buds are showing on landscape trees. Could it really be? Is spring approaching? We know that we'll have at least one more winter-like spell, perhaps even snow flurries - but the march towards green growth is in gear.

Over the winter, while we were stuck indoors, we cleaned and organized. You didn't? Well then, go ahead and get that done while you have a chance. We'll have plenty to do once the landscape recognizes that it is time to go on a growth spurt, making every attempt to overcome our puny efforts at controlling lawns, weeds, and hedge growth in our yards.

One of the problems I have this time of year is waking up at 3am thinking of the maintenance chores I need to do to be ready for summer. I realized that if I just put a checklist together for each season . . . I'd be ready, and wouldn't worry over what I might be forgetting. This does not mean that I am prepared or looking forward to the chores, but the checklist makes things easier. Here are some tips you might be able to use.

Sit down with a notebook and begin listing all the things you think you need to do. If family members are around, ask them to contribute ideas. After the first pass put the notebook aside, because later that day and the next week you'll think of other things. Write them down.

Your list won't look exactly like mine or your neighbor's, but we'll have some commonality. Once you begin keeping a real checklist, it will be easy to get a start next year. Here are some of the items on mine.

Check out the roof. Put a ladder up and look around. Get someone else to do this if you're uncomfortable with ladders and/or climbing on to your roof. Look for damage that might have occurred during our winter windstorms. Examine the vent boots for cracking. This is a prime area for the beginnings of a roof leak. Trim tree branches away from the roofline so bugs and animals don't have an easy time dropping down on to it. Check flashings and decide if the gutters need another clean-out.

Walk slowly around the house looking at the foundation, the windows, and the siding. Check foundation walls, floors, concrete, and masonry for cracking, heaving, or deterioration. A new crack in your concrete floor, slab or foundation may warrant a professional inspection. Make notes.

A section of your checklist should be allotted for equipment that needs maintenance. You only have to do this once, and you'll be ready for next year. My list includes buying oil and filters for yard equipment, string for trimmers, and charging and/or replacing batteries. Consider organizing your maintenance manuals in one area where you can check them quickly for maintenance specifications. Even better: write the specifications right on your checklist.

This is just the beginning. Go to the internet and type in "spring checklist." You'll be amazed at the volume of information. Take what works for your own list. Now, sit back and relax, knowing that you are really ready for next year's chores – and might not wake up at 3am thinking of them.

Tips and Tricks: Spring Cleaning for the Desk Bound

The "spring cleaning" tradition originated years ago in cultures and climes where the home was closed up throughout the winter, and the welcome arrival of spring meant throwing open windows and doors, taking furniture and rugs outside, and thoroughly cleaning every inch of the home, top to bottom.

Now, with cleaner heating fuels, our homes don't get as much "heavy dust" when closed up, and here in the mountains our wonderful climate gives us periodic breaks from the cold in which to let in the sunlight and air out the spaces on warmer days.

As many spring cleaning checklists circulate the news and how-to magazines this time of year, I think they are missing an important area: the home office. By home office I mean that place where we go to pay bills, search the internet, read a book, and file important papers.

If you dread the thought of filing and organizing your desk you are not alone. The only thing worse than looking forward to filing is looking forward to paying the bills or figuring out the taxes. If you enjoy these tactical organizational events, you are 1% of the earth's population or you might be from another planet.

I have come up with 8 tips and tricks for accomplishing desk organization with the least amount of complaining.

- Start with time and rewards. Most of us vastly underestimate the amount of time a task will take us, especially the ones we don't like. So, figure out how much time you think you will need to clean your office and then triple it. Then decide what you will give yourself for accomplishing the chore. Dinner out? A new book from Amazon?
- Clutter. Begin by clearing space on the floor. Yes, the floor. Then drop about eight or ten file folders on the floor, spread out. Now start with the top of the clutter pile and drop similar items – "2014 Taxes" on folders which you can label later.

- Toss stuff. Really. If it does not go on to a folder pile, throw it into a wastebasket pile. If you are one of those people who cannot throw anything out, put these things into a "misc." folder. After a year if you haven't touched it, throw it.
- Now that you've cleared the desk – wow – you forgot what color the desktop was! Use a damp towel to clean the surface.
- Near place for the critical. Sit at the desk and look around. Do you have enough cubby spaces or drawers to put the things you use all the time? If not, consider adding a small bookcase, set of drawers, simple bin, and/or file rack. Keep everything else farther away in bookcases, file drawers, or cubbies.
- Catch-all drawer. Yes, have one drawer where you throw things you don't know what to do with or things you don't want to get lost. When the drawer becomes un-closable it is time to sort it.
- Hangar hooks. Hang up your book bag, your keys, your jacket, whatever seems to end up on top of the desk that can be hung up. "Honey, where are the car keys?" "On the car keys hook." Oh.

Finally, add light. Most of our offices don't have enough light. Buy an LED or CFL bulb with a higher output than what you have. You'll feel as if the place is brand new.

Cleaning Tips and Tricks Part 1

If your family is like ours, no one is excited to have "cleaning" on the top of their to do list. It is easy to procrastinate because there are so many other things that we want to do. So, let's talk a little bit about actually keeping things clean as you go about your daily routine, and some tips and tricks on cleaning in general.

Here's a trick question: When is the best time to clean a shower? Answer: When you're still in it! Look at it this way: you're already all wet, the shower (or tub) is already all wet - why not keep a small brush in there with you, and a squeegee if your shower has glass panels - once a week brush up the surfaces at the end of your shower, squeegee, and then when you're done, towel the surfaces clean and shiny (yes, your towel then goes in the laundry).

You can perform a similar trick with your refrigerator - you know it's a pain to clean the refrigerator - everything comes out on to the counter, you are rushing to wash down the surfaces with the warm soapy water the manual tells you to use . . . the food is sitting out and getting warm . . . what a mess. Instead, every week or so, pick one small area to clean - remove the items just from that area and wipe down the interior - put the stuff back in, and that section is done. Rotate areas every week and your refrigerator will always be clean and neat. This technique also works for the freezer areas.

I don't know if the following works on bed bugs, but it does work on dust mites, which are teeming inside those "dust bunnies" rolling under your bed . . . in addition to weekly vacuuming and dusting, once a month you can load your comforters and pillows into the dryer and run them on as high a setting as the label says you can (and IF the label says you can) - this will roll all the dust and pollen and other "stuff" out your vent to the outside.

Do you have one bathroom sink drain in the house that clogs up about every 3 months? I'll bet this drain is the one you use most often - the one you wash your hands in after you put that shiny pomade on your hair (hair goes down the drain); the one that you rinse the comb in (hair goes down the drain); the one you brush your teeth over (toothpaste down the drain); the one you wash the bottom of your shoes in (yes, people do this); the one you floss over and sometimes the floss goes down the drain; you get the picture. Where does all this stuff go? Sometimes it does not leave the drain, rather it sits in what we call the "P Trap" . . . this is the U-shaped thing under the sink that takes what you put down the drain to the sewer, or, in our case, our septic tank. It is U-shaped because we want a little water sitting in the bottom of that "U" so that we don't have to smell what is in our septic tank. The only problem is, this water "trap" collects all these solids coming down the drain (including your wedding ring or earrings, OOPS!) and then creates a gummy gooey mess out of everything.

No problem, you say, I have a can of DRANO! But hold on. Even DRANO will have a hard time clearing this goop from your P Trap, and it's not good for your septic system or for your plumbing. So, here's how to clean your P Trap. This is also how to retrieve your rings and earrings when (not if) they fall down the drain.

If you look under your bathroom sink you will discover that some smart plumber figured out that we might need to actually remove or clean our P traps regularly. Look at the U-shaped pipe. If your home was built in the last 40 years, you'll see white PVC pipe. Now look at the fittings. These are actually designed to be removed by hand! No wrenches. Great. If you see anything else, or you see metal, then you need to regroup and find a professional.

Get a small bucket or towel, because the trap will have water in it. Now twist the fittings off - you should be able to turn them by hand - they have right hand threads just like a bottle top - envision looking down on the connection and your hand is at the top - right tightens and left loosens - the trap will drop off and you can now empty the gunk out of it, including the earrings you couldn't find. NOTE: If the fittings don't budge, you can use channel lock pliers gently on them. Make sure you're turning in the right direction!

Use a brush and some hot soapy water and clean out the trap. Now re-install - hand tight is fine - run some hot water down the drain and make sure you tightened the fittings enough - and you'll have a clean drain, a well running sink, and the smugness of knowing you can now retrieve your jewelry without calling a plumber.

NOTE: If this procedure did not unclog your sink drain, then guess what - time to consult a professional! Drain snakes and air blowers are the next solution tools. Drain cleaners should always be a last resort - in spite of what the bottle says, they are not good for your plumbing or for the environment.

Cleaning Tips and Maintenance Tricks Part 2

In Part I we talked about getting your cleaning chores into your daily routine, so that you wouldn't have to take the time to clean separately. Sometimes, however, we have to schedule cleaning chores. Let's talk about making these simpler so you can finisher earlier and get back to what you want to do. There may be people out there who LIKE to clean - if you're lucky enough to have that built into your personality - you can come visit my house tomorrow.

A Progress reader brought up a wonderful point about last week's article. I am going to call this "SCF", or "Small Child Factor". When I was talking about what goes down your bathroom drain and how to get it out, of course I forgot to mention the "SCF". This factor is like living on another planet entirely - the things that you will find in drains - crayons, small toys, parts of stuffed animals, Lincoln Logs, Legos, paper clips, pen caps, toothpaste (yeah! let's empty the whole tube down the drain!), and other strange liquids and solids, etc. - you almost need to have a trap door under there. This applies even more so to toilets - since the bowls are child height and obviously put there for every child's amusement flushing whatever they can find down the bowl. The "SCF" adds complexity and interest to otherwise boring household mysteries.

Back to the bathroom. Take a look at your toilet. In addition to the normal cleaning routine, you should inspect the inside of the tank periodically. The parts that enable the toilet to flush are mostly plastic and they start to deform over time. If you use bleach in your toilet or chlorine tabs in your toilet tank, the plastic parts will deteriorate in half the time because of the caustic nature of bleach. I know that bleach gives our bathrooms that fresh clean smell, but remember what we said previously about bleach and your septic system? Bleach slows down the waste breakdown process by killing the very bacteria that make this process work. It's a tradeoff - use a friendlier cleaner, like soft scrub, or use a solution of baking soda with water, or with vinegar, for the bowl. Sprinkle baking soda around the rim and use your toilet brush. This will both clean and deodorize your toilet. Then, if you really love that bleach smell, use a less concentrated bleach solution for the parts of the tank and bowl that won't be flushed into your septic tank.

Back to the parts inside the tank. Does your toilet "run", meaning, can you hear the sound of water continuing after the tank has filled? If so, the flapper could be hanging up, or the edges of the flapper might not be sealing. If you're handy, you can buy a kit at the hardware or home improvement store that will help you change out the tank flapper as well as the flush valve. Take a good look at what you have in the tank now so you can choose the right kit.

Now check the bowl itself. This is something I always do when I am inspecting plumbing on a job. Straddle the toilet bowl (face the tank) and GENTLY try to rock the bowl with your legs. Check for side to side movement and front to back movement. You can also kneel on the floor in front of the toilet and use your palms to see if there is movement. If it moves, then the bowl might not have been set properly, or the bolts holding it to the floor might be loose. About every 15th toilet I check is loose. If your toilet moves, try checking the bolts first. If the bowl moves front to back (rocks), you can put small pieces of plastic, or shims, under the areas that have the gaps. To make things look better and to secure the shims, you can then put a very small bead of caulk around the edges. Leave a couple of inches clear in the back in case your toilet starts leaking someday - you want to know about it. Use a small bead because getting a toilet base up from the floor for repair with a lot of caulking is difficult. Also, be aware that a rocking toilet over time will ruin the wax seal - this is the seal between the floor plumbing and the toilet bowl flange - and if this seal fails - you guessed it - you will look up and see your ceiling dripping water or your crawl-space will suddenly be very moist.

Save Time with these Cleaning Tricks

Do you have either a mental or written "to do" list that includes house cleaning? If you're like me, the answer is yes, I do have a list like this, but I keep putting the "clean" item back on the bottom of the list every time it starts to float to the top. If this sounds familiar, then these tips are for you.

First trick: fool yourself into cleaning when you're NOT cleaning. Instead of choosing a dedicated time to clean everything, clean things in the middle of doing other things. Here are some examples.

Clean the shower and tub when you're still in it. You're already all wet, the shower (or tub) is already all wet - why not keep a small brush in there with you, and a squeegee if your shower has glass panels - once a week brush up the surfaces at the end of your shower, rinse, squeegee, and then when you're done, towel the surfaces clean and shiny (yes, your towel then goes in the laundry).

The refrigerator - you know it's a pain to clean the refrigerator - everything comes out on to the counter, you are rushing to wash down the surfaces with warm soapy water . . . the food is sitting out getting warm . . . what a mess. Instead, pick one small area to clean in the refrigerator on the fly as you are taking something out. Remove the items just from that area and wipe down the interior - put the stuff back in, and that section is done. Rotate areas every week and your refrigerator will always be clean and neat. This technique also works for the freezer areas.

Dusting. This trick is similar to the refrigerator trick. When you are in your closet, keep a few rags hidden in the corner of a drawer, pull one out and spend 30 seconds dusting down 2 or 3 shelf areas. Then stop and go back to dressing or whatever you were doing. With this habit established over a few weeks, your closet areas will be dust free.

Along the "dusting" lines, I don't know if the following works on bed bugs, but it does work on dust mites, which are teeming inside those "dust bunnies" rolling under your bed . . . in addition to vacuuming and dusting, once a month you can load your comforters and pillows into the dryer and run them on as high a setting as the label says you can (and IF the label says you can) - this will roll all the dust and pollen and other "stuff" out your dryer vent to the outside.

You get the idea. Just keep rags or dusters handy and clean a small spot or area and rotate. This technique may even eliminate the need to have "cleaning day" at your home. If you train your family members to do the same – you'll be amazed at how much free time you now have for the other things that are far more fun than cleaning.

Don't Try This at Home

Did your teacher or parents ever say to you, "Don't believe everything you read"? Growing up, we tend to rely on the printed word as fact. In current times, we sometimes do the same thing with the Internet. We google something and think that everything we see is true because it's in the public view.

The other day I was researching cleaning techniques because one of my aims in life is to clean my house in the shortest possible time. If there are tricks and special techniques, I want to know about them. In my research, I came across collections of ideas that seemed odd. At first I thought that the writers were being humorous. Turns out that the ideas I am about to tell you are presented in all seriousness. One article is titled, Tips Your Mom Never Told You." The reason your mom never told you these tricks is because they don't work, take more time than they need to, are expensive, or are just plain silly. If you use any of these tricks to clean your house, I am sorry if I offend.

- Clean your bathtub with vodka. Right. Go to the ABC store, buy a pint of vodka, and pour it into your bathtub. Something tells me that buying a container of soft scrub is going to be cheaper and safer to use than 80 proof alcohol in the bathroom.
- Rid your couch of pet hair with packaging tape. Buy a roll of 2-inch tape, rip off about 4 inches, and set about wiping across every inch of your couch. After about 40 minutes of this activity about 5% of your couch will look wonderful. Trust me on this, use a vacuum cleaner with a brush.
- The other pet hair trick on the internet is taking a latex glove (like your doctor wears) and rubbing it back and forth quickly across the furniture fabric to build up static electricity. The static electricity will then attract the hair. About the only thing you will catch with this technique is a series of electric shocks and laughter from anyone watching.
- Use Kool Aid to clean your toilet. This probably works but even if it doesn't stain the bowl I expect that one container of soft scrub is going to be cheaper than 36 packages of Kool Aid.
- Use newspapers instead of paper towels to clean windows. Have you tried this? Maybe I was doing it incorrectly, but it was a mess. The newspaper ink was everywhere – on me, on my clothes, on the floor, but amazingly, not on the window. I would call this inconvenient and messy at best. My advice is to use a squeegee with one cotton side and one blade side with a mild detergent solution.
- Rid glassware of spots by running the dishwasher on empty. On empty? Right – no glasses in the wash, no spots. I thought that was what Rinse Aid was for?

- Improve air flow through vents in your home by waxing the grill. Really? I tried this and there was just as much dust after this waxing as there was before the waxing, not to mention the waste of time hand waxing each grill piece. Simply run a damp cloth over the grill from time to time and change the filter.

I did not make these up!

Take the Cleaning Challenge

Here are two questions that most people will give exactly the same answer to: "Are you a good driver?" Everyone will say yes. "Are you a good cleaner (when you have to clean)?" Nearly everyone will say yes. Are we born good drivers and good cleaners? We may have been lucky enough to get a formal class on driving in school, but I do not remember the cleaning class.

Can you pass this quiz?

Mixing the following two household chemicals together for cleaning can cause an explosion: A. Water and Windex B. Silver Polish and Borax C. Bleach and Dish Detergent D. Bleach and Ammonia. If you answered D, you are correct! Not only is there a risk for explosion when bleach and ammonia are mixed, they will also produce deadly fumes.

Speaking of bleach, why is it not a good product to use for cleaning sinks and drains? A. The smell can be overwhelming B. Bleach is expensive C. Bleach will stain porcelain D. Bleach will interrupt the functioning of your septic system. If you answered D, you are correct. You should use bleach very sparingly, if at all, if you have a septic system. The bleach kills bacteria, both good and bad, and bacteria are what help your septic system break down waste materials.

Your LCD TV screen should be cleaned with: A. Windex and a paper towel B. 409 and a cotton cloth C. Straight ammonia and a squeegee D. a clean cotton cloth and plain water. If you answered D, you are correct. LCD screens are best cleaned the way you would clean your eyeglasses; no harsh cleaners, and a soft cloth. Paper towels can actually be abrasive, and leave lint behind.

How much "HE" or High Efficiency detergent should you put into a full laundry load in a high efficiency clothes washer? A. One half cup B. One cup C. Three-fourths of a cup D. An eighth to a quarter of a cup. If you answered D, you are right. Most of us put about twice as much detergent into our wash than we need. To see for yourself, put the detergent amount in you normally do, and when the load is done, run it through another rinse cycle and take a look in the drum. See all the suds? Not only will you be save money by using less detergent, your clothes will come out brighter and cleaner.

Area rugs are best cleaned by: A. Turning upside down when the top gets dirty B. Rubbing them with a damp cloth C. Put them in the clothes washer with the regular load D. Shake them outside and then vacuum. If you answered D, you are right again. Putting them in the clothes washer will work, as long as they are on a gentle cycle apart from the regular wash, and if the label says they are washable. Otherwise, your entire load of clothes will have tiny little pieces of rubber backing material on them. This debris from rugs is devilish to remove. The best practice is to shake them outside, and if possible, run a vacuum over them to get the rest of the soil lifted.

So you passed the challenge! Next week we'll finish up the discussion with myths about cleaning. Now it's time to take a break from cleaning and go outside in this lovely spring weather in the mountains.

Four House Cleaning Myths

Spring is springing, and even though we're experiencing some lingering cool temperatures, summer will be here before you know it. You are not going to want to be inside, cleaning your house, so let's talk about the fastest way to get those ugly chores behind you so that you can go outside and play.

Myth One. Use cold water for all of your laundry. Not true. At least, not true all the time. Your instincts on this one are good; go ahead and use hot water for items that can stand it (read the label) and have the potential to retain bacteria, mold, viruses, and dust mites, such as bedding and towels. Use a hot dry cycle on these items also to finish getting rid of mites and really sanitize them.

Myth Two. Use old newspapers to wash your windows. I'm not sure who came up with this idea, but if you've tried it, you found out it was not pleasant, and probably not effective. In addition to the newsprint coming off all over your hands and staining the frames, wet newspaper crumbles, gets caught in the corners, streaks, and leaves residue. Instead, use a squeegee with a half and half solution of Windex and water, or a half and half solution of white vinegar and water. Ingles sells a clever squeegee with a cotton cover that works great and cuts window cleaning time in half.

Myth Three. To clean your fireplace box at the end of the winter season, vacuum the ashes up. Have you tried this? You might as well throw out your vacuum cleaner after doing this if you tried it; the ashes are so fine that they get into everything, including oozing out the sides of your vacuum bag and getting caught inside the hoses. Instead: take a large trash bag and put the open edge near the ash pile. Wearing a pair of disposable gloves and using the edge of a throwaway magazine or a piece of cardboard, slowly and carefully transfer the ashes into the bag. Then use a dustpan and brush to get the rest. Don't use a damp cloth or paper towel in this operation, as you'll only embed dirt into the box floor.

Myth Four. Put coffee grinds in the garbage disposal to clean and freshen it. Where in the world did this tip come from? Apparently from plumbers and handypersons who are waiting for you to call and tell them that your plumbing system is clogged up. The one time I tried this, I didn't even find the "smells nice" claim was true – it smelled awful. A much better way to freshen up your disposal is to keep a dedicated old dishwashing brush under the sink. When your disposal starts to give off that garbage aroma, pour some of your dishwashing liquid on this brush and put it into the disposal (not running of course!) and scrub the sides and the blades. Rinse thoroughly, running cold water through the unit. To freshen up, throw a small lime or lemon rind in with some ice cubes to sharpen the blades, and you're done.

More Creative Cleaning Tips

Readers write: "Please publish more interesting cleaning tips." I am not sure why people like these articles. But since I enjoy either inventing or researching ways to reduce your cleaning time at home, I'll share more tricks with you today. There are many products that can be used in unconventional ways to save you time.

RAIN-X. Rain-X is typically used on car and truck windows to make rain bead up and improve visibility. You can also use RAIN-X on glass or laminate shower doors. Apply it after you've cleaned the surfaces, and then expect to go three times as long before the next cleaning. This silicone based chemical will repel dirt as well as water. If you have a window in your home that seems to get dirty more often than the others, apply Rain-X inside and out and watch it sparkle for months. Use a slightly damp towel to polish the window or glass doors to remove any streaking after application.

OLD SOCKS. I bet you've already thought up great uses for old socks, but have you used them dampened to clean dust off plant leaves? Get dust out of corners? Use them to collect miniature dust bunnies out of the nooks and crannies of stairs that are hard to vacuum, and put them on a ruler to reach under cabinets. You can also use these for window blinds and polishing fan blades and fan light bulbs.

LINT ROLLER. Sold to remove lint and hair from clothing, these sticky tape rollers are great for getting pet hair and dirt in general off furniture. Also use them on wall corners where your dog or cat rubs its fur.

CAR POLISH. Car polish will work well cleaning gold, chrome, and stainless steel jewelry. The one I've found that works best is Nu Finish Car Polish. Take an old rag and apply the polish, rubbing the surfaces briskly. Before this dries, take an old toothbrush and put a few drops of dish washing soap on the bristles. Put the jewelry under the faucet with lukewarm water, pull out, and use the toothbrush to scrub the jewelry. It will foam up – rinse this off and scrub again. Make sure all the polish is off. Use a cotton towel to dry to a bright shine.

PAM. Pam, the no stick spray coating for pots and pans, can be used to spray candlesticks before you place the candles in them so they'll come out easily later, spray door hinges to stop squeaks, and on the grill of your car to make bugs stop sticking to it between washes.

PAINTBRUSH. Use a slightly dampened paintbrush in one hand and the extension attachment on your vacuum in the other to remove dust from bookcases, tops of cabinets, fan blades, and any other area where you know dust lurks. This can work better than just using the brush attachment on your vacuum, which tends to move the dust clumps around on the bristles.

Everything You Didn't Want to Know About Cleaning Your Bathroom

Just what you wanted to read about today – bathrooms. Added up over a lifetime, we spend over 15% of our waking hours in bathrooms. We all want shiny clean bathrooms to spend all this time in, but who has the time to clean them? Last week I read several how-to articles on bathroom cleaning. I stopped reading after the first one instructed me to mix up a complicated five ingredient mixture for the shower walls, a different mixture for the mildew, and another mixture for the drains. I just wasn't up for this level of complexity.

Would you like some time saving tricks? Before these methods will work you have to get the grunge off. After an initial deep clean, you should not have to do it again if you keep up with it. Deep cleaning is a lot of work and we all hate it, but go ahead and get it done and then use the tips that follow to keep things fresh. Unless you just arrived from Mars you probably already have your favorite cleaning chemicals, whether home-made or store bought.

- Each bathroom should have its own supplies. No sense in running from room to room looking for things.
- Keep another washcloth in the shower and use it once a week to soap up the walls and then rinse. It only takes seconds. Use a pump bottle of liquid antibacterial hand soap. Leave this in the shower for when you get ambitious.
- As you prepare to exit the tub, use a washcloth to apply soapy water to the sides of the tub as it drains. Rinse with clean water. As the water finally leaves (wow, where did all that come from?), soap up the bottom also, and rinse.
- Buy a removable toilet seat. These assemblies simply lift off the base, and make it fast and easy to clean that tough area between the tank and the back of the lid.
- Wipe and dry out the sink when you're finished. This takes just seconds, and will keep the sink and fixtures looking like new. This trick works well with the kitchen sink too.
- A squeegee is a real timesaver, but an even better one is to apply Rain-X to your glass shower door (but do not get Rain-X on plastic, vinyl, or aluminum). Yes, the water will bead up and run off just like it does on your car's windshield, leaving

the surface clean. You can also use Rain-X Anti-Fog to keep mirrors clear.

- Use an automotive polish for acrylic surfaces to achieve the same clean beading as Rain-X does on the glass.
- Use one of the liquid soaps when you bathe instead of bar (cake) soap. Many bar soaps leave a residue on surfaces.

These are just a few of the things you can do to keep things perpetually clean and shiny without expending much effort. If you have small children (or big ones for that matter) in the house, demonstrate these techniques to them as well to reduce your own workload.

March Cleaning Madness: Fast Cleaning Tricks

It feels like we have less time the longer we navigate life. Why is that? The to-do list gets longer, and we continually underestimate how long a project is going to take. This may not be one of your problems, but it is definitely one of mine. The optimist in me is always saying, "Oh, the deck stain job won't take more than a weekend," only to discover it took 4 weekends.

To counteract this woeful inability to plan my time, I came up with ways to shorten the time it takes to do the things that I don't want to do. Cleaning, for instance. If you are someone who lives to clean, don't read any farther. If you are like me and you'd like to spend your time tinkering, building, planting, or working on your favorite hobby, then read on.

The whole idea here is to perform as many cleaning chores as possible WHILE you are doing other everyday activities. This way, you'll get double duty out of your available time. Here are some examples. After reading these, I know you'll be able to come up with more just like them. When you do, send them to me for the next cleaning article!

- Small bites. Clean the refrigerator by taking 30 seconds just before you reach in for something to moisten a cloth with a bit of soap and water and then do a small wipe down, just in that area. Then get on with your day. If you do this twice a week, your refrigerator will never need cleaning. This also addresses the problem of dirt and food sitting and congealing into a non-removable mass over a period of weeks on interior surfaces.
- Dusting. There are two great opportunities to remove surface dirt and dust. First, when you are collecting laundry to be done, take one of the items that isn't particularly dirty, like a cotton towel, moisten with a little water, and look around for the nearest TV. TV screens should always be cleaned with lint free cotton and water, never with cleaners. Use the towel to wipe the screen, then turn the towel to a dry area to remove any remaining moisture. Back into the laundry basket. One less thing to do. The second opportunity is in the laundry room. Take the same towel that is already moist and wipe down your washer and dryer and surrounding surfaces. One less thing!
- Microwave interior. When you are doing the dishes, whether by hand or in the dishwasher, open the microwave and remove the rotating table and table rest. Clean the table. At the same time, wipe the inside of the microwave and when your rotating table is clean, that's one less thing to do.
- Clean as you go. If you're in the shower cleaning yourself . . . take a few minutes to wipe down the interior of the shower. When you're emptying the wastebaskets, remove the plastic liner from the first one and then empty the others in to it. Won't fit? Have plenty of plastic liners in the bottoms of all your wastebaskets. Grab a second and keep going. You'll end up with 2 instead of 8 bags at the trash bin.

You get the idea. Look around as you are doing your chores to see what you can combine. These tricks will save you time and then you can spend that time doing what you love!

How to Organize Your Garage

How does your garage look right now? With spring spruce-ups leading to repair chores, the garage is fix-it central, with more and more items landing there for attention. Some garages are no longer garages – the vehicles have been put outside because there is no room for them inside.

Even banishing the cars and trucks to the outside doesn't help organize what you have. There is a law which states that the more cubic space there is in your garage, the more items will be attracted to the open areas of this space, until there is no space for humans to move between items. This is a corollary of the Closet Law, which states that it does not matter how large your closets are, the items placed in them will expand to completely fill the space.

The idea of organizing your garage may be so repelling to you that you are going to turn the page right now. But if you are an organizer and love having things put away, here are some garage organization tips.

The first key to getting garage space back is storing or hanging things on the ceiling or walls. Go online to get ideas by typing "garage storage" in Amazon.com. You'll see dozens of clever ideas that you can make yourself or buy. These organizers include racks that mount above your garage door, clever brackets that allow you to hang your bicycle(s) on the wall, and fixtures for hanging up wheelbarrows, weed eaters, and mops and brooms. Special pegboards are available that will get lots of stuff off the floor yet keep it handy for use.

The second key is a combination of open and enclosed shelving. Cabinets range from inexpensive plastic to durable metal. Choose the best one for your budget. Put the items you don't use as much in the enclosed cabinets and the stuff you use a lot on the open shelves where you can see them.

The third key is the workbench. If you already have a workbench, clear everything off it. Look at the area at the back by the wall – could you add a shelf to place bins or cans to hold small items? Does it need a pegboard? Also examine the area underneath the bench to see if you need storage there as well. After these simple upgrades, organize all the "stuff" that came off the bench back on to it. Theoretically, you now have bench space showing.

The fourth key is tool storage. Put a tool roll cabinet on your wish list. Harbor Freight and Global Industrial sell durable tool cabinets for less than $150. Having a nice cabinet will prompt you to put the tools away after using them, and allow you to roll your tools to the job.

The last piece of advice is optional but will really add pizzazz to your garage. If the floor material is concrete, why not paint it? If it's wood, why not refinish or renovate? And if the floor is dirt, consider some specialty gravel to dress it up. Next week I'll discuss how to choose and apply concrete paint.

Should You Get Your Ducts Cleaned?

In the last column, I talked about home improvement scams. Are duct cleaning services a home improvement scam?

To be complicated, the answer is yes, and no. It is not necessary to have the interior of ducts cleaned to have clean air, unless there are unusual or special circumstances. Special circumstances include seeing mold growth inside your vents, discovering that animals have made your ductwork their home, or seeing particles of dirt or dust actually coming out of your air vents. As a home inspector I can tell you that these three circumstances are highly unusual and rare.

If you do think you need the ductwork cleaned, then it is critical that you choose the right contractor. Duct cleaning done unprofessionally and without the right equipment will make the problem much worse and can actually damage your heating and cooling system. To choose the right professional for duct cleaning and HVAC cleaning services, go to the NADCA (National Air Duct Cleaning Association) web site: www.nadca.com. You should also read up on the topic by going to the EPA site: http://www.epa.gov/iaq/pubs/airduct.html#choosing. Duct cleaning services can be expensive, so it is worth the research time.

If you have read this far and decided that you do not need duct cleaning, and never want to clean them, here are several things that you can do yourself.

Keep the heating and cooling system clean. Preventing "junk" from getting into the ducts in the first place will head off the need to clean them in the future. To keep the system clean, check the filters every month. Vacuum the air vents and registers once a month. Register covers in the floor pull right out for this easy task. On the others, use a brush to dislodge the dust off the vent vanes. Get the heating and cooling system serviced twice a year – once in the spring, and once in the fall. This will keep all of the associated filters, coils, lines, and drains clean and open.

Inspect the ductwork where you can see it. You can do this yourself or you can talk someone else into going into your attic or crawlspace to look at the ductwork. Turn the system on with the fan running. You are specifically looking for leaks in the system. You will hear these and see these. I have been on numerous home inspections where I discovered missing duct tape, holes in the ductwork, and even disconnected sections of ductwork. If air is escaping from your ductwork, your electricity bill will be higher, the system will run more often trying to make up for the loss of compression. The unfiltered air that is entering through the leak will contaminate the ducts and your house air. If you or the person you talked into looking finds a leak, it is time to call a heating and cooling service tech to inspect and seal the ducts.

A tight duct system with properly filtered air will mean you do not have worry about getting your ducts cleaned, and will lower your heating and cooling bill.

Tips and Tricks for Healthy Air

In past columns, I've written about mold prevention in your home. Excess humidity and even small mold problems in your house may produce air that smells musty and contain mold spores, producing allergic reactions. Here is a collection of tips to improve your air quality year-round.

Find leaks. We've had a lot of rain this winter – make sure water isn't coming into your home in a place you're unaware of. Check all ceilings, baseboards, the attic, and the basement for stains or outright leaks. If you discover a leak, get the area repaired as soon as possible. Water that sits on interior surfaces for more than a day or two will damage the material and produce mildew and eventual mold.

Circulate the air. Create adequate air circulation by providing a return path for air to move back to the source. If all the doors are closed and your home does not have air return vents (often over the doors), the air will apply back pressure to the air handler or furnace and reduce the efficiency of the equipment. The reduction in air flow will increase humidity and stale air, creating opportunities for mildew.

In summer set your air conditioning fan to AUTO. If the fan is set to ON, condensed moisture on the condenser coil (outside) may not have enough time to evaporate, and end up back in your home's air.

Air Filters. Change the air filters in your heating and cooling system before they get cruddy. When they get cruddy, guess what is being blown back into the air you are breathing? I like to check them every

30 days. Buy the cheapest ones you can find and spray them front and back with filter spray. This improves their ability to trap and hold dirt and allergens. You can find filter spray in your local home improvement store.

If you have floor supply vents in your home, have you ever looked inside them? The grills are usually not fastened to the floor and you can lift them off. Vacuum these out periodically. As a home inspector, I routinely look inside these vents, and about a third of the time they still contain debris from construction from dozens of years earlier. In some cases I found them completely clogged. I've also discovered everything from petrified mice to Lego pieces under the grills.

Closets. Closets can be a perfect environment for fungi and mildew. Air vents are normally not supplied to these areas. If you smell any mustiness in your closets, try leaving the light on for a while (make sure the bulb is protected and not touching anything – you don't want a fire), which will raise the temperature and help dry the air. Open the door and air closets out periodically. You can also buy desiccant sacks to hang in closets to lower the humidity. They work much better than air freshener, which simply masks the mildew smell.

Use real plants. Real plants produce oxygen. Who couldn't use a little more oxygen? The downside is that real plants require care and feeding, and may raise the humidity slightly, which actually may be helpful in the dead of winter when we are suffering from dry air.

There are dozens of good ideas to keep your indoor air clean and healthy. Try a few of these and take a big healthy breath.

10

Home Maintenance Mistakes

Avoid These Home Maintenance Mistakes

You have your home maintenance checklist and you can't wait for the weekend because you love working on this list, right? If you are like me, then yes. But 97% of the population is not like me. 97% of the population is NOT thinking about maintenance chores, they are thinking about the fun they will have at the lake, on the trail, on the golf course, at the track, or at the party.

If you're not excited about household maintenance, you're more likely to short cut them. With this is mind, here are the four mistakes I see all the time as a home inspector. Making these mistakes will shorten equipment life.

- Using WD-40 for everything. It's great to have one can that can do everything, right? I mean, the ads for WD-40 say it has thousands of uses. This is true; it's a great product, but it should not be used for EVERYTHING. WD-40 is a solvent, with Naphtha Petroleum as its primary ingredient. Instructions include, "cleans, clears, removes, and loosens." So instead of roaming the house with a can, stop and think about the things that need heavier, or different, lube. Two examples are your garage door – chain and rollers – that should be treated with real (gooey) grease, and sticking door locks which should be treated with a graphite lube.

- Changing heating and cooling system air filters rarely and/or putting them in backwards. These filters are cheap insurance against dirty air, they improve the efficiency of the system when they are clean, and they are designed to go one way. Look for the arrow and make sure they point towards the direction of air flow.
- Not inspecting the condensate line from your air conditioning handler before summer. When your air conditioner fires back up in the spring months, the condensate line takes the water that the unit removes from the air and transports it to the outside as a liquid. Most of the time this line exits somewhere in your yard. Find this end before the line backs up and shuts off your AC. As a preventative against clogs, pour a little bleach or vinegar in the line at the handler every month or two during the summer. And if it does clog, you can take your wet-dry vac into the yard and suck the clog out. Trick: put a clear four-inch-long section (plastic tubing with clamps) in the line leaving the air handler so you can see if it's free running or not.
- Not getting regular service on your water heater. Let's assume you already have your water heater insulated and the temperature set high enough to avoid bacterial growth (122 degrees F.). The heater is going to work just fine for the first 2-4 years, but gradually sediment is going to form in the tank and the sacrificial anode rod is going to do its job by disappearing. In year four or five, suddenly you hear, "Hey! No hot water!" Keep your water heater going for 7-15 years by inspecting and draining it every year or two.

If you want to be really ahead of the game, research and buy what you need ahead of time (like grease for the garage door chain) and place it in an organized spot so you can find it when you need it. Add these to your checklist with notes and times, and the phone numbers of your favorite service people, and then off to the lake.

Household Chemistry Controversies

I need to tell you up front that this article may be controversial and I'll get some calls and emails from folks telling me that I'm wrong. But bear with me and keep in mind that even the professionals in a trade often disagree on how something should be done, or what products to use. The following advice is culled from dozens of articles, books, and interviews with plumbers, electricians, contractors, and home advice professionals. Let's talk about three household chemicals: septic system additives, anti-bacterial soaps, and the great standby, vinegar.

Last month when I wrote about septic systems, I received mail about my advice to not use additives. These readers asked if they should stop using them.

Septic system additives consist of enzymes which facilitate the bacterial breakdown of sewage. Because the human body uses enzymes in its metabolic processes, these chemicals are already plentiful in sewage. As a result, additives are not necessary for the adequate decomposition of waste materials in a septic system. In some cases, additives can even be harmful to the system by breaking down solid waste in the tank. These smaller particles then move into the drain field where they clog the soil and damage the system. Also, using enzyme or other chemical additives in your septic system will not lengthen the time between pump-outs. For the skeptics, google: "Do you need to add enzymes to septic tanks?" You will find both points of view, but the pro-additive advice is from companies selling the additives, and the con advice is from scientists and university studies. You decide.

Triclosan. We've been led to believe that bacteria and viruses can only be eliminated by using anti-bacterial products. Much of this preys on our fear of illness. One particular chemical, called Triclosan, is inviting scrutiny because recent scientific studies have reported that this chemical affects hormone balance and muscle function in animals. Triclosan is an ingredient added to many consumer products to reduce or prevent bacterial contamination. It may be found in products such as clothing, kitchenware, furniture, and toys. It also may be added to antibacterial soaps and body washes, toothpastes, and some cosmetics. The FDA (U.S. Food and Drug Administration) expresses concern on its own website over Triclosan, but stops short of saying that we shouldn't use it. Of particular interest to me is that Ingles has removed Triclosan from its store brand pump soaps. So, this is another area where you will need to make up your own mind. The fact remains that good hand washing with plain soap and water will do the same job that the anti-bacterial soaps do.

Finally, the old standby vinegar. We know that vinegar is one of the most versatile and ubiquitous ingredients in any kitchen. But, is vinegar an all-purpose household cleaner? The facts are that vinegar does work well for things like cleaning mildew in the bathroom and cleaning and descaling the coffee maker. It has anti-bacterial properties as a rinse (I am not enamored of the smell). But, vinegar is not really effective at dirt and grime removal. This includes counter tops and sinks. My advice here is to be selective with your cleaners and not rely on just one product. You decide!

11

Emergencies and Staying Safe In Your Home

Are You Prepared for an Emergency?

Recent stories about people being trapped in their cars in the icy road conditions got me thinking about emergency supplies for the home. Do you have an emergency stash of food and other items that could help you in the event of a storm? Here in the mountains folks are more prepared than elsewhere because we know conditions can change quickly. But the national statistics for preparedness show that only about 20% of households have an emergency "kit."

"Honey, where's the flashlight?"

"Under the kitchen sink."

"I found it. It doesn't work!"

Sound familiar? Do we check our supplies before needing them? Probably not.

Here's a checklist for supplies and gadgets you should consider keeping in your home somewhere easy to find. Add your own ideas to this starter list and check functionality once or twice a year.

Working flashlight. Notice I said "working." Over the last 10 years battery technology has improved substantially, and battery life, which used to be awful, is now a year or more. Purchase the best flashlight you can afford. You won't be sorry when you need it. Add them to areas where they are handy: kitchen, bath, and bedroom.

First Aid Kit. If your home is like mine, Band-Aids and ointment are scattered between bathrooms and the kitchen. For emergencies don't spend time hunting supplies down. Assemble a small box with Band-Aids of various sizes, antiseptic wipes, gauze pads, ointments, etc. Add to this a small supply of critical medications. Your doctor will probably write an extra prescription so that you have some for the kit. Keep this entire kit separate from the other supplies so items don't go missing. Check expirations on medications when you do your inventory once or twice a year. You can also purchase ready to go (store) first aid kits locally or online.

Tool bucket. Another cry also heard in our house is, "Honey, where is the hammer?" Do tools somehow get separated from their proper place in your home? It does not matter how many tools you have, when you need one of the six tape measures you own, they have all transported to another planet for the duration of the need. Then they appear, one by one. To make sure tools are available in an emergency, buy a bucket and place critical items in it. Then put this bucket in a closet and tell everyone that it is off limits for everyday use.

Cell phone backup power supply. If your power goes out you will be glad you bought this inexpensive little device. Half the size of a pack of cards, this unit charges via USB and holds the charge for a year.

Hand crank radio. For about $35 you can buy a radio that includes a 12-volt charger that will keep you abreast of the news and weather, albeit not without some hand and arm exercise. If you purchase an inverter for your cell phone you can also keep mobile communications going.

Add your own items to this checklist and be ready. Murphy's Law states that if you are completely prepared, nothing will happen!

How to Hold a Successful Emergency Drill

Remember when you were in school and the fire drill would go off? We were trained on exactly what to do when that alarm sounded. Don't spend time grabbing your belongings, move fast, be orderly, go to the rendezvous point to be counted. Fortunately, the drills I remember didn't include any real fires or other catastrophes. But they easily could have been real, and we were ready.

Are you ready? What if a fire broke out in your home – would everyone know what to do? Sure, I know the point is to get out as fast possible, but a little preparation for this event would make it smoother and faster, and could possibly save a life. Although I am talking about family drills, it's just as important if you live by yourself. A well-executed evacuation can mean the difference between life and death.

The key to handling emergencies with calm action is preparation and practice. Here is a preparation checklist to review with yourself and your family. Have a meeting to discuss and confirm the answers to these questions, and then hold drills.

- What emergencies might you face that would cause you to leave the house? List these. They could include a tornado that has hit your house, a fire that has broken out in the house, a fire in the woods next to your house, or a flood.
- Map escape routes for each area of the home. Actually walk these off to see how long it takes.
- Assemble a backpack or bag to place in a convenient area of each floor of the home with some basic supplies. These could include a first aid kit, a few tools, water, flashlight, etc. This bag can be as simple or extensive as you like. Go online and look up bug out bag and you'll get some ideas.
- Do you need a fire escape rope ladder in the second story? These are about $35 each. If you save lives, the cost is well worth it. Get one for every high deck that does not have egress, and one for any upper story rooms that have only one way out.

Usually the one way out is down a staircase and fires tend to move up these drafty areas first.

- Do you have fire extinguishers on hand in areas of the home? When I was inspecting homes, I saw the correct extinguishers in the right places about 3% of the time. The Kidde FA110 Multi-Purpose 1A10BC is only $20 from Amazon or a local home improvement store. You should have an extinguisher on every level and within reasonable reach – about 30 feet.

Once you have these details covered, you should actually practice getting out of the house. Never used a rope ladder? You don't have to actually go down the ladder, but do get the ladder out of its packaging and see how it attaches to the railing or the window. Speed counts.

Run the drill numerous times and if you have children, make a game out of it. Discuss possible scenarios and ask how family members would handle them.

We hope the time never comes when we have to carry out our evacuation plan but if it does we will be prepared and as safe as we can possibly be.

Averting Household Catastrophes

We've all been there. "Honey, the water heater has flooded the basement. Honey, the clothes washer is leaking. Honey, the circuit breaker keeps tripping. Honey, the Christmas tree just caught fire. Honey, the lint in the clothes dryer is smoking."

Well, maybe you haven't had these things happen to you. Good! You're lucky or smart, or both.

Ten years ago, a neighbor two streets over from us left to sail on a two-week cruise. One week after they left we all noticed a small pond forming in front of their home. When a neighbor went up to the home to investigate, he was shocked to see water running out under the front door and down the steps. He smartly turned off the water supply to the home and called the owners. The discovery: the owners had not turned off the water supply to the inside of the home, and a toilet refill valve failed from age, spraying an upstairs bath with water at five gallons per minute. Although homeowners' insurance covered the gutting and rebuilding of the home's interior, many special items were lost, and the owners had to move out of the home for eight months.

None of us want to be surprised by something this awful. Here's a short checklist for the big things that could cause you to have an instant bad day.

Flooding. When you leave for more than a few days, turn the water supply off. Make it a checklist item. This doesn't mean that a pipe can't burst while you're at work, but it tempts Murphy's Law a lot less. Purchase six or seven small battery powered water leak detectors. Place these under the kitchen sink, behind toilets, behind the washing machine, at the water heater, and other places where you want to know right away if there's a leak. Third, inspect your washing machine hoses every six months and replace them if they are cracked or feel squishy (technical term).

Fire. Fires are devastating and happen fast. Lower your chances of having a fire by making sure the electrical system is in code and not overloaded (an electrician can inspect it for you), installing dual sensor (photoelectric/ionization) fire alarms throughout your house, and testing them every year. Have fire extinguishers handy in the kitchen and in the garage, and have an emergency escape ladder on the second floor if there's no alternate staircase.

Critter invasion. An assortment of animals can get into your house and die inside walls, leaving the house uninhabitable for a long time. To prevent this, make sure crawlspace and eve vents are securely screened. A bat can climb into a hole the size of your fingertip. Wasps and yellow jackets can also find their way in through a tiny crevice. Take a thoughtful walk around your home and look for areas of entry.

Patch and caulk as needed. Termites can enter the home hidden from view and cause very serious damage. Get a termite professional inspection every year.

Mold. In a high humidity environment mold grows very fast and can overcome an entire floor. If this proceeds over a summer season unchecked, the only remedy will be professional remediation. This is expensive. Don't let this happen to you by making sure the highest humidity in your home is 60% or less. Check all areas, particularly basement rooms. Simply installing a humidifier can keep mold away entirely.

Don't count on luck – think ahead! You'll be glad you did.

How to Choose Backup Power for Your House

Does it seem to you that the weather patterns have exhibited some strangeness over the last few years? If so, you are not imagining it. Folks may argue over the cause, but it's a fact that things are changing. One of the results of current weather patterns seems to be more frequent and more powerful storms. We are lucky here in the mountains to have more stable weather than other parts of the earth, but you never know when this might change.

Are you a "prepper"? Not to be confused with "preppie," the high style of prep school students, to be a prepper means to prepare for bad things that might happen, such as tornadoes, hurricanes, and power grid failures. Preppers are part survivalist and part planners. I am not a survivalist (the reality TV is too scary), but I think I'm part prepper. When I lived in Florida this meant I installed a whole-house generator with three 100-gallon propane tanks after two hurricanes in a row took our power out for more than a week each time.

The power outages seemed fun for a time. But after 20 minutes without the internet, and thinking about the meat in the freezer, the fun wore off. After I installed the whole house generator, the internet was still out but I had a working refrigerator, air conditioner, and washing machine, and a long extension cord to run to the neighbor's house.

Seriously, have you thought about power backup? Some readers have already made their plans, and are at the ready with a variety of power generators. North Carolina is not the same as Florida of course, and typically when our power goes off it doesn't stay off long. But if you are one of those "you never know" people and want to know more, here is a quick guide to keeping the meat frozen.

There are two ways to go on a power backup plan. The difference is price and convenience, and to some degree, safety. The least expensive route is purchasing a small portable generator. These units come in all shapes and sizes, and deserve your research effort to make sure you get what you want and need. A great place to find this information is Amazon.com and ConsumerReports.org.

Drawbacks to using portable generators include having to keep the unit ready to go, running power cords to appliances, having to keep the unit outside (never run a portable generator inside your home), putting up with the noise, and they require a gasoline supply. Advantages to portable generators include minimal set up time and low cost to purchase. They can be transported in your truck to other areas, and no connection to your house power is necessary (or allowed).

Portable generators can present hazards, such as fire danger in storing the fuel, inadvertently having fumes enter your home in the form of carbon monoxide, and explosion danger when re-fueling a hot engine. Never try to connect your portable generator to house power for convenience. Your portable could back feed the power line and electrocute a worker repairing the lines.

In the next column, I'll talk about the other form of power backup – whole house generators.

Choosing Backup Power: Whole House Generators

In the last column, I talked about choosing a backup power source for your home in case of power outages. While this area of the country has not seen the catastrophic weather events the other areas of the United States have, this doesn't mean it cannot happen here. If you've thought about a power backup solution for your home, here are the things to consider.

In the last column, I mentioned that the least expensive, but also the least convenient, generator is the portable type. Remember that a portable generator must be run outside the house and that means extension cords running to all of the things you want to power. At some point this becomes impractical. Enter the whole house generator solution.

Whole house generators are systems that kick in automatically 30 seconds after a power failure. They use a special transfer switch attached to your power panel to automatically isolate the generated power from the incoming power for safety. The Kohler Company recently ran an ad on TV showing a family having a party, playing music, and dancing. The power goes off. The house goes dark for about 5 seconds; the generator outside turns on, and the family resumes partying. The experience "in real life" is not far from this situation. If you need, or want, instant backup power without any hassle, a whole house generator is the way to go. But do not let the name "whole house generator" fool you. Most of these systems will run about 60% of your household circuits, and not central heating and cooling, without spending considerably more money.

Advantages of installing a whole house generator include almost no maintenance, the unit kicks on whether you are home or not (nice security feature), they are quieter than portables, they run on propane which does not "spoil" over time like gasoline does, and you can choose what appliances and tools you want to automatically power. The unit "exercises" itself for 15 minutes every week and needs an oil change only every two or three years.

You are thinking, "This is too good to be true." You guessed it: the only real downside is money. These systems can be expensive to install. You will need a propane tank, a plumber to install the lines, and an electrician to install the transfer panel inside the home. Add from $700 to $1500 to the price of the generator for the professional installation. You can do some of the work yourself but you should employ qualified folks for the electrical and gas lines.

Have I discouraged you? I hope not. View the variety of choices on Amazon, and check out the internet for whole house generator power calculators. If you only want to run your refrigerator, lights, small appliances, TVs, tools, water heater, and a portable air conditioner or heater, prices before installation start at $1,800. If you want to run the entire home, including central air and 220V appliances, you'll need to dig deeper into your wallet. But the first time the power goes out and comes back on from your generator 30 seconds later, you'll be pleased.

Checking Your Home for Water Leaks

I got a call from a friend who lives in Jasper, Georgia. She was beside herself as she told me about arriving home after a few days visiting the North Carolina coast only to find water running out of the front door of her little vacation cabin. The valve inside a toilet broke, and she had not thought to off the water to the cabin before she left.

I listened empathetically to my friend's story. Although not on the same scale, this happened to me in the form of a gasket failure under a toilet in a home we were selling. Murphy's Law! The upstairs bathroom was located directly over the family room. It was pure luck that I was home that day moving furniture to observe the growing circle of wet drywall and dripping water from above. As I ran to shut off the valve at the toilet, I considered the awful damage that could have been done if I had not been there.

We are fortunate here in the climate and topology of the mountains to have our own personal or community wells. The other day I heard someone say, "I don't worry about water drips, I have a well." This shocked me, since I grew up in a family that was very conscious of conserving resources whether it was water, electricity, or zip lock bags. Did your mom ever say, "Turn off the lights, you're not in the room!" or "Shut the door, do you think we live in a barn?"

Here are some ideas for keeping floods out of your home and reducing water waste.

- Know where your water shutoff is. Make sure everyone in the house knows where it is. This way, in an emergency, you can cut it off quickly.
- Inspect all of your water using appliances. Are there shutoffs nearby? Most appliances - especially toilets and faucets - have their own shutoff valve so that maintenance can be done. In an emergency, you can turn the water off locally without running for the main shutoff. These local valves themselves can leak, so look at everything.
- Check bathroom shower heads where they meet the arm coming out of the wall. Turn on the shower and look at where the water is running – is any water running into cracks in tile or behind the shower arm or tub faucet? Check tile thoroughly - small cracks can develop over time and allow water to run in under the shower pan.
- Check the toilets. Take the toilet tank cover off and listen. 90% of the time, if you have a leak in the toilet internals, you will hear it. It might be a slight hissing, or just the sound of intermittent drips. Non-leaking toilets should be completely silent. If you are unsure, put a few drops of food coloring in the tank (not the bowl). If the color shows up in the bowl, your flapper valve at the bottom of the tank is leaking. These are easy to replace and cost just a few dollars.

Add water aware checks to your spring and summer walk-around, and consider turning off the main supply in your home when you leave for vacation. A few extra minutes will help keep Mr. Murphy away on vacation too.

Winter Heating Safety Tips

It seems like winter arrived in the mountains suddenly, as if someone flipped the "winter" switch to "on." Switching from air conditioning to heating can present surprises, like when it doesn't work and it's 22 degrees outside.

Our expectation is that the heating equipment will work perfectly the first time in over 7 or 8 month of being idle when we hit that thermostat, but this isn't always the case. It's sort of like driving on tires when we've not checked the air pressure - we don't really notice diminished performance over time. Or, we may find on the first cold day that the heat does not work at all. This is one of the "things your mother told you but you didn't want to hear" sayings: regular and preventative maintenance pays off over time. You can go year to year and the appliances continue to work, but if you have a plan to service them regularly, they are less likely to surprise you with a malfunction.

Here's how to keep your heating appliances working and keep you safe.

Fireplace. Most of us throw the wood in there and light it. That's it. Simple. But wait! Before your next fire, take a look at a couple of things. First, is the damper working properly? Is it closing all the way? Opening all the way? If it's not, you're sending warm/cold air up the flue when you're not using the fireplace, and when you are using it, it may not be venting properly. Take your fingers and press at the rear of the firebrick inside the fireplace - is there any give, or any loose brickwork? This is important, as loose brickwork can cause the area behind it to receive more heat than it should and cause damage. Open the damper and look up into the flue, or use a digital camera and point it up the flue and take a picture. Is there a buildup of creosote, a fire byproduct? If you discover any problems. call a professional chimney sweep for an inspection and a cleaning. Creosote buildup can cause a fire where you do not want it - inside the chimney.

Gas Logs. At the beginning of the heating season, have your gas company inspect the gas lines and valves. Try to negotiate this service into the purchase of your propane. As a home inspector I always look at gas lines that are visible and do the "sniff" test - get your nose right up to the valve and see if you can detect any sulfur, or "rotten egg" smell. Propane is a heavier than air, odorless gas that has the smell added to it so that we can recognize a leak. I check the outside storage tank valve too. About 8% of the time I find a leaking line or valve. This is serious - a spark or flame can cause propane in the air to explode. If you do think you have a leak, call your propane company immediately and exit your home; do not turn any electrical light switches on or off as you leave. They can generate a spark that could ignite the gas.

Gas Furnace. If you have a gas furnace, follow the tips above for leak detection. What I'm about to say here now also applies to gas log and wood burning fireplaces: these appliances can produce small amounts of carbon monoxide if there is a leak, or if venting is not adequate. Both propane gas and carbon monoxide gas settle to the floor, so I recommend that you add a carbon monoxide detector near floor level if you don't have them installed in your home now. Get them at any home improvement store and follow the directions for installation.

How to Keep Your Water Supply from Freezing at the Well

Most of us here is the mountains get our water from a well rather than a municipal supply. This introduces a set of maintenance chores that some folks don't know about, especially people moving here from homes that had a city water supply. The first time the pipes freeze at the well there is consternation and confusion. One Progress reader wrote, "We had just moved to a small cabin in Shooting Creek. I woke up on a very cold Saturday morning in January hearing 'Uh oh, no water,' from my wife. I had no idea what to do. Luckily there was a handyman next door and he placed a small heater in the pump house."

We have been lucky to not suffer the extremes that other locations around the planet have endured, from extraordinarily cold winters to flooding to raging fires. But that does not mean that we're not going to get some of these effects in the form of wild weather swings in the coming years.

Winter water supplies can freeze and cause problems throughout the plumbing system. Unless you turn off the heat inside the house without draining the pipes, the problems will be contained to the lines from your well to your house, the equipment in the pump house (or under your unrealistic looking well rock). Also unlikely but certainly possible is that the water line from the well was not buried deeply enough in the ground or soil was removed so that there is not enough depth for the line to stay unfrozen.

Typically, though, water freezes at the well head. This is the one area above ground that, if not insulated and protected, will freeze if we have many days in a row of very cold temperatures. Here's what to do if your water freezes at the well.

First, check to see if there is any insulation around the pump equipment if you have a pump house. If not, add an insulation box. Next, add a heat source. One way you can do this is to purchase two lamps. Install them near the pump equipment with 60 watt bulbs. Do not use CFL – CFL bulbs don't convert electricity to heat like incandescents do. You can install a thermostat switch that will turn on your lamps when the temperatures drop below 40 degrees. Or, when winter arrives, simply turn on the lamps. Why 2 lamps? In case one burns out.

Some folks use an electric space heater in their pump shed. I'm not enamored of this solution, because the thermostats tend to be unreliable, and it's a potential fire hazard.

A far more elegant and energy efficient solution, and one that works really well under those unrealistic looking rocks, is heat tape. This solution is available as a kit if you are feeling handy and want to do the job yourself, or you can simply call your well drilling company and ask them to "heat tape winterize my well."

The heat kit is composed of a length of electric heater cable with a power light at the plug and a built-in thermostat. Follow the directions carefully wrapping the pipes, and place the thermostat in the coldest area. Then insulate the pipes or purchase an insulation kit for your rock so that the heat is contained. The thermostat will turn the heat kit on when the temperature drops below 40 degrees. Voila! May you never hear the words, "Uh oh, no water!" again.

Electrical Safety - What is GFCI?

Here's one of those "what is this and why do I need it" questions - what is GFCI? Take this quiz: GFCI stands for: A. Garden and Farming Collective Insurance B. Garage Floor Cement Improvement System C. Ground Fault Circuit Interrupter D. Girder Framing Continuous I-Beam.

Did you answer C.? You're right! A GFCI, or Ground Fault Circuit Interrupter, is a special electrical receptacle that shuts off the electricity at that outlet if, for example, you drop your hair dryer in a bathtub full of water (hopefully you are not in the tub at the time). Any imbalance in electrical current will be sensed by a GFCI outlet and will make it trip very quickly. Hundreds of electrocutions are prevented by these simple and relatively inexpensive devices. However, as a home inspector, I see many homes that do not have any GFCI protected outlets. Is your home one of them?

GFCI protection was not included in building codes until 1973 when GFCI protected outlets were required outside. Now, GFCI outlets are called for in kitchens, bathrooms, garages, unfinished basements, crawlspaces, and laundry rooms! So, depending upon when your home was built, you might or might not have them in some or in all of these places.

Because GFCI protected electrical outlets are so effective at protecting you from fatal electrical shock, you should consider installing these special receptacles in the places where you do not have them now - kitchen, bathrooms, garage, and laundry room. Can you do this yourself? Although many people are comfortable with turning off the power to the outlet and following the directions in the GFCI box carefully, I would recommend hiring a licensed local electrician to do this work. These special outlets are not expensive, and for the electrician, they are a very easy job to install.

You should test your GFCI outlets periodically. Here's how: you will see two buttons on the outlet, usually right in the middle between the two plug areas. Plug a lamp or night light into one of the outlets and then push the "TEST" button. The outlet's "RESET" button should pop out, and the light that you plugged in will go out. If the "RESET" button pops out but the light does not go out, the GFCI outlet hasn't been wired correctly. If the "RESET" button does not pop out, the GFCI is defective and should be replaced. If the outlet is working correctly, all you need to do is push in the "RESET" button and power will be restored to the outlet. If the "RESET" button will not reset, this, along with the other problems I just mentioned, should be repaired or replaced by a qualified person (electrician).

Safety Tips for the Holidays I

Can you believe it? The holidays are here! Why is it when you are a small child Christmas and your birthday take 48 months to roll around again? Or so it seems. Now, as adults, your birthday and the holidays take 4 months to roll around again. Or so it seems!

The holidays are a good time to talk about safety in your home. With the rush on, so much to do, and guests arriving, it's easy to overlook small things that can make a difference to our health and well-being. Here are some tips to keep you and your loved ones safe. Let's begin with the kitchen.

The kitchen is a wonderful place with coziness and mouthwatering aromas, and the kitchen is definitely the place to be during the holidays, with food tasting, treats, warmth, and good cheer. The kitchen can also be dangerous. Most home fires start in the kitchen, most people who show up in emergency rooms with serious cuts get them in the kitchen, and nearly all burns are a result of an accident in the kitchen. So, consider the following tips:

Be sensible with knives. Keep them sharp so they don't slip, and keep them in a holder, not loose in drawers. If you drop one, let it go! Don't pull the knife towards you when cutting - if it slips it may cut you. Cut away from you, and think while you're cutting. Most of us reading this have had a careless moment we'd rather not repeat. In spite of what you read, humans really can only concentrate on one task at a time (try telling that to a texting teenager). In the kitchen, we have to be careful with everything that is going on and keep our focus.

Fires. Keep a fire extinguisher somewhere handy. If you don't have one, go to one of the home improvement stores to get an all-purpose extinguisher. Murphy's Law says if you have one, you won't need it. Watch what you're cooking to make sure it doesn't overheat or get forgotten. When you place a hot pot on the counter, put a potholder over it or its utensil to let others know it's still hot. Verse your children in kitchen safety, especially the hot pots and surfaces.

The range. All ranges manufactured in the last 18 years are required to have anti-tip brackets installed. These are metal devices designed to prevent freestanding ranges from tipping forward. They are normally attached to a rear leg of the range or screwed into the wall behind the range, and are included in all installation kits. A unit that is not equipped with an anti-tip bracket may fall forward if enough weight is applied to its open door, such as that from a large Thanksgiving turkey, or a small child. A falling range can crush, scald, or burn anyone who is caught under it. As a home inspector, this is one of the items I check for on my inspections. And guess what? 80% of the time these brackets are missing. I attribute this to the installer not wanting to go to the extra effort to put this bracket in. To find out if your range is missing a bracket, simply grasp the rear top panel area of the range and try to tip it towards you. It should not tip up. If your range is missing a bracket, you can buy one in any home improvement store. If your home is fairly new, look through all the "stuff" the builder gave you when you moved in - the bracket might still be sitting in a plastic pouch with the instruction manual! If you're handy, you can install this yourself.

Electrical cords. Unplug appliances in the kitchen when you're not using them. Check the cords for fraying. Don't overload the circuits. Don't run extension cords from somewhere else in the house into the kitchen - yes, people do this - they can trip someone and knock what its powering over, spilling contents. It can also overload circuits, raising the potential for fire. Be careful using appliances near the sink. If something you're operating falls into water, step away and let the ground fault circuit cut the power to the outlet powering the appliance. Check to see if your kitchen has GFCI protected outlets - if you are not sure what these are, ask someone to help you. These circuits have been required in kitchens since 1987 and will save your life if you drop an electrical appliance into a sink of water and you're still touching it.

Safety Tips for the Holidays 2 - Lighting

In the movie National Lampoon's Christmas Vacation, Chevy Chase puts up the family holiday lights with much effort and fanfare. There is a scene where one of the relatives goes into the pantry to retrieve an item for the kitchen preparation and we see that very thing that we know we should never do: 22 electrical cords plugged into 8 plug extenders all plugged into one receptacle outlet! It looks ridiculous and we all find it hilarious when grandma flips the switch off to the room and of course everything connected to the receptacle goes off, completely dumbfounding Chevy and everyone standing outside looking at the grand display.

We would never do this, right? I mean, we would never plug a bunch of cords into plug extenders and extension cords and run them all over the house and yard, right? Ah, but we do! We have all done it. Luckily, our electrical panel decides enough is enough and a breaker trips somewhere. Or at least we hope it will.

As much as I do not want to dampen your wild "Chevy Chase" holiday spirit, here are some tips and tricks for dealing with the holiday lighting chore.

You plan your outside holiday lighting, right? You sit down and draw out where the lights are going and what you will need for extension cords, junctions, etc. You read and follow the directions for how many strings of lights can be hooked up end to end. You go to the basement or attic and pull the neatly stored lighting . . . NOT! This is not what happens at all. Here's what happens - you go to the box overflowing with last year's light sets, some working, some not working, burned out bulbs and all - everything thrown together in a pile in a box not labeled and all the extensions cords and extenders are missing because someone in the family needed them and knew they were in this box in January. Whoops! So much for getting the lights up in four hours or less.

The trick to all this is organization and forethought, but we all find this difficult to do when we have so many chores this time of year. A better way: this year, sit down at the table and PLAN where things are going. Review where you are plugging the strings in and make sure your outside outlets are GFCI protected - these outlets will cut the electricity if water shorts the circuits. Do not use indoor extension cords for outside use - they are not designed for this use and can heat up and start a fire. Use cords that are labeled for outdoor use, and make sure they are heavy enough for the run you are planning. How do you know this? The label on the cord should help you figure this out, or ask a handy person who can add up the wattage of what you are installing and help you choose the right cord. The longer the run, the heavier the cord should be. Read the directions that came with the light strings to determine how many you can safely connect together. Assemble a kit as you go about your work this year - separate boxes that have the cords, with a copy of the plan you just made attached to it. When you take the lights down, re-check the function and serviceability of the lights and fix them NOW before you store them away. Then, store them in a place and with labeling so that the contents do not get raided during the year. When next year arrives, you will be so pleased with yourself!

Do the same with your inside lights - plan the runs and connections, and make sure you protect the cords from vacuum cleaners and people and pets tripping over them. Small area rugs or duct tape help in this regard, but be careful that the cords do not run under furniture feet or other objects that can apply pressure to the cord and damage it.

Safety Tips for the Holidays 3 - Heating

This week I'll talk about some hazards that are common at this time of year as we fire up the furnace, the fireplace, and the gas logs with the first round of freezing temperatures as we sail into the holidays.

Our expectation is that the heating equipment will work perfectly the first time in over 7 or 8 month of being idle when we hit that thermostat, but this isn't always the case. It's sort of like driving on tires when we've not checked the air pressure - we don't really notice diminished performance over time. Or, we may find on the first cold day that the heat does not work at all. This is one of the "things your mother told you but you didn't want to hear" sayings: regular and preventative maintenance pays off over time. You can go year to year and the appliances continue to work, but if you have a plan to service them regularly, they are less likely to surprise you with a malfunction.

Here's how to keep your heating appliances working and keep you safe.

Fireplace. Most of us throw the wood in there and light it. That's it. Simple. But wait! Before your next fire, take a look at a couple of things. First, is the damper working properly? Is it closing all the way? Opening all the way? If it's not, you're sending warm/cold air up the flue when you're not using the fireplace, and when you are using it, it may not be venting properly. Take your fingers and press at the rear of the firebrick inside the fireplace - is there any give, or any loose brickwork? This is important, as loose brickwork can cause the area behind it to receive more heat than it should and cause damage. Open the damper and look up into the flue, or use a digital camera and point it up the flue and take a picture. Is there a buildup of creosote, a fire byproduct? If you discover any problems. call a professional chimney sweep for an inspection and a cleaning. Creosote buildup can cause a fire where you do not want it - inside the chimney.

Gas Logs. At the beginning of the heating season, have your gas company inspect the gas lines and valves. Try to negotiate this service into the purchase of your propane. As a home inspector I always look at gas lines that are visible and do the "sniff" test - get your nose right up to the valve and see if you can detect any sulfur, or "rotten egg" smell. Propane is a heavier than air, odorless gas that has the smell added to it so that we can recognize a leak. I check the outside storage tank valve too. About 8% of the time I find a leaking line or valve. This is serious - a spark or flame can cause propane in the air to explode. If you do think you have a leak, call your propane company immediately and exit your home; do not turn any electrical light switches on or off as you leave. They can generate a spark that could ignite the gas.

Gas Furnace. If you have a gas furnace, follow the tips above for leak detection. What I'm about to say here now also applies to gas log and wood burning fireplaces: these appliances can produce small amounts of carbon monoxide if there is a leak, or if venting is not adequate. Both propane gas and carbon monoxide gas settle to the floor, so I recommend that you add a carbon monoxide detector near floor level if you don't have them installed in your home now. Get them at any home improvement store and follow the directions for installation.

When Was Your Last Fire Safety Inspection? Part 1 of 3

It is unfortunate that our area has witnessed severe home fires recently. We read about this regularly in the newspaper every single year. Many of these fires are so powerful that the home burns completely to the foundation and both people and pets are injured or lost. The emotional trauma that follows is doubly damaging.

As a home inspector, I find situations and conditions that could lead to a home fire, and I list them prominently in my report for the buyer or seller. However, you may not be buying or selling a house, so let's talk about some things you can do right now keep your home and your family safe.

You know you should have smoke (fire) alarms in your home and test them once a month, right? Be honest – are you testing them? I find that these alarms are either not working or missing batteries 50% of the time. I also find that 70% of the homes I inspect do not have enough smoke detectors. What is a few hundred dollars compared to losing everything? For $200 you can buy 4 modern detectors that wirelessly communicate, and if there is a fire in the basement, you will know about it instantly in your upstairs bedroom. In my opinion you can't have enough smoke detectors. Make sure you follow the directions for locating these carefully to minimize false alarms.

Do you have a fire escape plan? Here in the mountains we may have a bedroom that is a long way from the ground and you may have to exit a fire through a window. Put a plan together with your family that includes materials, such as a rope ladder. Do you remember fire drills in school? "Oh yay, get to get out of class for 15 minutes!" you exclaimed. There was a good reason for these besides disrupting your class – in a real fire you'd respond fast. You should have the same procedure at home right up to and perhaps even including a drill. If you have children, you can introduce some fun and rewards for fast times, knowing what to do, etc.

Most fire victims die from smoke inhalation. In your fire escape plan, consider including some inexpensive see through smoke hoods – available on Amazon for less than $10 each. People panic in smoke and fire, and being able to breathe, see, and move fast are important.

Do not turn spaces in your home into sleeping areas if they are not considered bedrooms. The reason for this is that bedrooms always have a second exit – either a door to the outside or a window to the outside. The window should be at least 24 by 20 inches, and not more than 44 inches off the floor so people can get out without difficulty.

Kitchen fires represent a big piece of the statistical pie. Buy a fire extinguisher for the kitchen area and keep it 10 feet away from the range in the exit pathway so you can reach it. Don't wear loose clothing when cooking, and don't wander away or get distracted when you are cooking. Does your range have the anti-tip bracket installed? Try to pull the range toward you from the back. If it tips toward you, have a qualified person install the bracket, which should have come with the unit.

When Was Your Last Fire Safety Inspection? Part 2 of 3

Last week we talked about fire safety in your home. Most of the time electricity is our friend, but it can become dangerous quickly when we stretch cords carrying it or overload circuits. On one home inspection, I climbed the stairs to the second-floor hallway and followed a yellow extension cord plugged into a hall receptacle all the way down to the end of the hall to a bedroom where the cord wrapped around a bookcase, disappeared under the carpet and then re-emerged to run under the bed and then across the floor to a window air conditioner. The cord was very hot to the touch - the air conditioner was running, laboring to suck enough electricity out of this little cord to operate. The longer and smaller the wire, the more difficult to transmit power – hence the heat in the cord. This is an electrical fire waiting to happen. I turned to the homeowner who was following me and asked her to turn off the window unit until we could figure out how to run power to it from a nearby outlet. "We only have two outlets in this room and they are all used," said the homeowner. I explained, "Your window air conditioner draws a lot of electricity. This hot extension cord could set fire to anything it contacts – the carpet or drapes, for example, and the rest of the room would follow. The cord on the air conditioner says right here on the label that it should not be plugged in to an extension cord for this reason." I showed her the label on the cord. "Wow, I had no idea," she said, "I'll call an electrician and have them install an outlet for it!"

Look around your house right now and make sure extension cords you are using are the modern type with grounded outlets; make sure they are not running under furniture or carpets, and make sure they are not overloaded with appliances or lamps plugged into them. How can you tell if they are overloaded? Use common sense. If you have a 2-outlet receptacle on the wall and one cord is plugged into the top, and an extension cord runs from the bottom, and there are 3 things plugged into that, it's too much! Feel the cords. They can be warm, but never hot.

Take a look at your wall switches and receptacles and make sure they are not loose where they attach to the wall. If you want to know if the wiring is correct (properly grounded, correct polarity) in each one of them you can purchase a circuit tester for a few dollars from any home improvement store. If you discover that you have incorrectly wired receptacles, an electrician should be called to correct it.

Check your lamp sockets for the recommended wattage and don't put a 100 watt bulb into a socket rated for 60 watts – it could overheat. With the newer high efficiency bulbs this is less of a concern, as these bulbs run much cooler.

Next week – fire safety in your laundry and workshop.

When Was Your Last Fire Safety Inspection? Part 3 of 3

Last week we talked about preventing fires in your home. This week let's inspect your laundry room and workshop.

When was the last time you looked behind your clothes dryer? Come on, now, really? As an inspector, I get to look behind clothes washers and dryers all the time, and it's not a pretty sight. Sometimes I wish I had time to hook up my mini-vac and suck out all that lint and dust from the floor. On a good day, I will find the dryer vent unobstructed, not kinked or crushed, and running properly to the outside of the home without lint blocking the outside vent fins or screen. If I do find any of these problems, I make sure to flag them in the report, as clothes dryer fires account for over 12,000 residential home fires a year.

These fires start when the dryer overheats because lint is blocking the vent or has built up inside the dryer. To prevent this from happening to you, take your vacuum and clear all the lint and dust from behind your dryer, and then check the vent on the outside of your home. If you see any clogs or partial clogs, suck them out with a vacuum and check regularly from then on. If you can't find your vent on the outside of the house, then you have a larger problem; your dryer vent could be running to your attic or your crawlspace. These are never good places to vent your dryer, as the chances of fire go up dramatically when you can't check and clean the vent, or a buildup of lint is created anywhere inside the house, particularly a hot attic in the summer.

If you find that this is the situation in your home, get a qualified building contractor or handyperson to run the vent outside. This is probably not a job you want to get involved with, as the vent run might have to go through walls or to the soffit area of your attic.

Don't leave home when your dryer is running. Yes, it's inconvenient, but if a fire starts, you'll smell it first and can take action. Use common sense with what you put into the dryer – read labels carefully to make sure the item is not a synthetic that could burn. Clean the lint from the dryer's internal screen every time you put a load in to minimize lint getting into the other parts of the machine and building up.

Practice fire safety in your garage or workshop. Have a portable fire extinguisher handy; use care when storing flammable liquids, and don't fill your hot lawn mower inside the workshop or garage when it runs dry. Liquids can spill on hot parts of the machine and ignite. Leave the machine outside, let it cool (time for a cool drink), then fill it.

Gasoline and other flammable liquids should be stored in a shed or weatherproof box outside the garage. Make sure tops and lids are tight and the containers are well labeled. When was the last time you poured gas into your 2-cycle weed eater thinking you'd added the oil to it, but you hadn't?

The same rules apply even more so with electrical safety in garages and workshops – don't overload circuits and be careful using extension cords.

When you're done with chores and projects, take the time to put things away and check that machines are off and unplugged, and fuels are safely stored. If you are a smoker, you should be extra vigilant and not smoke around flammables in your shop. Have a safe place to extinguish your cigarettes.

Don't Become a Fire Statistic over the Holidays

According to the United States Fire Administration, fires in the home during the holidays kill an average of 400 people, injure 1,650 more, and cause nearly a billion dollars in damage annually. The opportunity for home fires goes up dramatically during this time of year – don't become one of the statistics. Here are some tips to keep you safe.

Make sure you have a fire extinguisher on each level of your home and that the extinguisher is not expired. A flash fire will not give you time to run to the garage or shed! Check your smoke detectors – push the red button to test – if you have any that are not working, you should have it repaired or buy a new one. Smoke detectors are not expensive, and you can even add wireless detectors to your home if you need additional units. Placing a unit in every bedroom and every hallway is a good idea if you do not have them in these locations already. Have an escape plan, and if you are unsuccessful putting out a fire with an extinguisher, get out pronto and let the professionals handle it.

The most likely sources of unintended fire are your live Christmas tree and your Christmas lights. If we combine a tree that is dry from lack of water with an extension cord that is worn, frayed, or overloaded, or with hot bulbs, we have a recipe for fire.

Keep your tree watered, and don't leave the lights on when you're not in the room. Every season when you drag that tangle of cord and light strings out of the attic or basement, do you check for frayed wires, bare spots, gaps in the insulation, broken or cracked sockets, and excessive kinking or wear? Decorating the tree is exciting, but if you do the inspection pre-work, you'll have less opportunity for problems later on as you go about the fun stuff.

Remember the scene in National Lampoon's Christmas Vacation in which Chevy Chase had hundreds of light strings throughout (and on top of) the house, and they all plugged into one outlet in the pantry via "extenders" all plugged together? I'm sure your holiday arrangement does not look like this, but if it's similar, you might want to consider reviewing the instructions attached to each string set for the maximum that can be linked together. If any of your light string cords or your extension cords is hot to the touch, you have too many devices on one line.

A great new light set available these days is LED. This stands for "Light Emitting Diode". LED solid bulb lights consume about 20 times less energy than incandescent lights do, they are durable because the bulbs do not have filaments, they do not contain mercury, and they run cool, which is a great safety feature. There's only one problem . . . they are expensive! A string of 50 Christmas LED lights cost about $15 versus a dollar or two for the traditional set. Over time, however, the cost of producing these great fire resistant lights is expected to go down. Put this on your Santa list for next year. Enjoy the holidays and stay safe.

Could Your Appliances Catch Fire?

Could your home appliances catch on fire? Do you know if any of your appliances are under a recall notice?

Most of us answer, "I don't have a clue about whether any of my appliances have a recall on them. How would I find out, anyway?"

In past columns, I've talked about fire safety related mostly to clothes dryers. This is the first appliance most people think about, since safety instructions, home inspectors, contractors, and service people remind us constantly that excess lint in our dryer line can catch fire. What you might not know is that clothes dryers are actually in second place – and a distant second place - when it comes to the numbers of fires caused in the home and the number under recall by manufacturers.

Clothes dryers caused about 13,800 fires in homes over the last 10 years with over 8,000 recalls on units, but RANGES caused over 44,000 fires and had nearly 17,000 recalls! Many other appliances, from toaster ovens to dishwashers to microwave ovens catch fire as well, and are under a variety of recall notices from manufacturers. According to Consumer Reports, more than 15 million appliances have been recalled in the past five years for defects that could cause a fire in your home.

Here's what you can do to check out your current appliances and protect yourself when you buy new ones.

Go to www.recalls.gov and www.saferproducts.gov to see if any of your appliances are listed in the databases. If they are, you will see manufacturer information which you can use to decide what to do.

Did you get a home inspection before you bought your current home? Your home inspector may have checked the appliances against the recall lists. If so, it would be listed in the report you received.

When you purchase new appliances, register ownership by filling out the card that came with it. Are you one of those people who throws all this stuff away in the trash after you unpack the appliance? Think again next time – if this appliance that you just plugged in is under a recall due to a defect that could cause your home to catch fire – wouldn't you want to know about it? You don't have to fill out those little postcards either – you can go to the manufacturer's web site to register, or call them. When we buy a new car, we expect to be notified quickly if there is a potential problem with our vehicle – consider doing the same thing with your appliances.

Consider keeping a fire extinguisher handy in your kitchen, in your laundry room, and in your garage. Sure, you might not be home when the fire starts, but why not have this extra insurance? The second part of this strategy is making sure your smoke alarms are all working and that you have them in the right places. No smoke alarms?

Two-thirds of home fire deaths occur in homes without them. Smoke alarms are inexpensive insurance for you and your loved ones.

Home appliances are getting more complex as electronic ignitions are added to gas appliances, touch pads are replacing push buttons, and "intelligent" circuit boards are built into refrigerators, ranges, and dishwashers. More complexity means more opportunity for error. Checking for recalls and registering your appliances will go a long way in keeping you and your family safe.

Protect Your Home from Fire

The recent news about out of control fires in the western areas of the U.S. is alarming. Draught, heat, and lightning contributed to the loss of homes and lives in the Arizona mountains. While we are fortunate to not have the same conditions here, there are things that we can do to reduce the chances of suffering a fire in our home or a wildfire near our home. Talking to Clay County residents who have experienced a fire in or around their homes brings to light the devastating toll that a fire extracts. Here are five things to consider in keeping your home and family as safe as possible.

Wildfires come from a combination of events that can be weather related (lightning) or human error (a brush pile burn gets out of control). To help minimize wildfire from reaching your house, create a buffer zone. Clear out excess vegetation and underbrush growing next to your home. If you are on a mountain slope, clear the areas below the house, as wind will tend to drive fire upslope. If you start a burn pile, choose the right weather conditions, check permit requirements and safety regulations (www.ncforestservice.gov), and maintain a watch at all times. Most importantly, leave your home if a fire is approaching and allow professionals to fight it. This is not an easy thing to do, but your life is more important than your possessions.

Fires inside the home develop when circumstances allow an ignition source to meet something flammable. This can be an electrical spark caused by faulty wiring, overloaded electrical circuits producing excess heat, a short (conductors touching) in an extension cord, and flammable materials (fuels, thinners, adhesives) igniting. Kitchen fires, candles left unattended, and curious children with matches are also in the top ten house fire causes. Here are some things you should do to help prevent these situations.

Install early warning devices throughout your home. This includes fire detectors that have dual sensors - ionization and photoelectric smoke sensors – to give you maximum warning, and carbon monoxide detectors. Don't use cost as an excuse – combination fire detectors are available from Amazon.com or your local hardware store for about $25.

Have a plan. Decide ahead of time what you and your family will do if the alarms go off and/or a family member shouts "Fire!" Several times a year run a fire drill. Practice the drill until you are clear about where to go and what to take with you. Consider investing a few dollars in smoke hoods ($10) to keep in bedrooms. Smoke inhalation can be more dangerous than heat. If your home has a second story, make sure there is a way to get to the ground. Fire safety ladders are about $35.

Educate your children about the dangers of fire. Give them responsibility during drills. Ask them to help develop a checklist that you can all use to audit the safety of your home and the overall preparedness of the family. Rotate leadership and tasks and reward with positive reinforcement. Use the internet to locate additional checklist items.

Does this sound like a lot of work? It is. But the payoff will be increased peace of mind and an improved margin of safety for you and your family. Just ask anyone who has escaped from a burning house.

Top Five Causes for House Fires

Several weeks ago, two homes in Clay County burned to the ground. A house fire is devastating. Everything is lost to a fire; most importantly, lives can be lost in house fires. The National Fire Protection Association (NFPA) reports that in 2014 there were 1.3 million fires reported in the United States. Of these fires, 494,000 were house structure fires, causing 3,275 deaths, 15,775 injuries, and $9.8 billion in property damage.

There are things you can do to help avoid a fire in your home. Here are the top five causes for house fires and what you can do to prevent them. Remember, from the moment an uncontrolled fire starts to the point where the structure is fully engulfed is usually less than five minutes. It is critical for you and your family to get out of the home as quickly as possible and not try to put out a large fire yourself.

1. Cooking. The top cause for house fires is walking away from the stove when you've got things cooking on it. Be extra vigilant when cooking – don't leave "just for a minute" – it's too easy to forget that you have something on the range. This does not apply to crockpots, which are designed to be left unattended.
2. Heating. Number two on the list are fires produced from space heaters and fireplaces. Over 80% of heating equipment fires are caused by a space heater catching something next to it on fire, or creosote buildup in a chimney catching fire. The rest are an assortment of strange things that folks do to keep warm, like bringing their barbeque grill inside (don't do this). Don't leave space heaters unattended, and keep them away from flammables such as curtains. If you have a wood burning fireplace, have it inspected and cleaned annually.
3. Smoking. Third on the list is falling asleep with a cigarette burning or accidentally dropping a burning cigarette in the house. It will smolder and then catch whatever it fell on on fire. I know the smokers will be upset with me – but I am going to say that

it's a much safer practice to smoke outside. Use sturdy, deep ashtrays. Never smoke in bed, or around supplemental oxygen.

4. Electrical and Appliances. The fourth cause of home fires is faulty electrical equipment, electrical appliances, and wiring. This includes excessive use of extension cords, lint catching fire in your clothes dryer, and frayed wire arcing. The NFPA reports that wiring distribution, including lighting, is the cause of 65% of electrical fires. To reduce the chances of an electrical fire, have your home wiring inspected periodically by a licensed electrician or home inspector, don't overload circuits, and supervise the use of appliances (meaning don't run to Ingles when your dryer is running).
5. Candles and Christmas Lights. The fifth cause of house fires can be almost completely eliminated by only having candles burning when you are in the same room and keeping them away from flammables. Inspect your Christmas lights for problems and wire fraying. Turn lights off inside when you leave the house or go to bed. Don't overload circuits or use too many extension cords. Feel the cords to make sure they are not hot.

Finally, purchase and keep a small fire extinguisher where it can be easily reached in your kitchen, in your bedroom, and in your garage. Check operation of your smoke detectors. If you have decks or upper story bedrooms, purchase an escape ladder and keep it handy.

Follow these tips and be safe.

What You Should Know About Fire Alarms

Several readers have written to me since I wrote about the top five causes of house fires, asking about smoke detectors. What type should they buy, and are older detectors ok?

There are three types of smoke alarms available – ionization alarms which have been around since the 70s; photoelectric alarms which were first patented in 1972 and gained market ground in the 1990s but did not sell as well because they were more expensive. The third type is a combination of ionization and photoelectric technologies.

What's the difference? Ionization smoke detectors are designed to sense fast flaming and fast moving fires. This type of detector will go off quickly in the high heat of a flash fire and a fire that has relatively little smoke associated with it. A photoelectric smoke detector detects visible and invisible particles, as you would have in a fire that has barely begun, but is producing smoke. An example of this would be a cigarette falling into bedding and smoldering.

According to the National Fire Protection Association (NFPA), non-working detectors exist in 20% of the homes with alarms installed. As a home inspector, I can tell you from experience that this statistic is on the low side. Every third home I inspected had non-working fire detectors. Over half of the nation's residential fire deaths happen in homes without any detectors at all, and the overall risk of dying in a home without working smoke detectors is twice as high as the risk in homes with working detectors.

What to do? First, if you already have detectors installed, take a look at them. Write down the identifying information and look up the type (ionization or photoelectric, or both) on the internet or take the information to your favorite home improvement store so that they can identify it for you. If you discover that your detectors are only the ionization type, I recommend that you purchase additional detectors of the photoelectric type to supplement your current detectors. Follow the directions for usage, testing, and maintenance. This includes periodic testing (test button), replacing batteries annually, and keeping vents clear. You can purchase detectors that are completely battery operated and communicate wirelessly with all the others so that when one goes off, they all go off.

To give you an idea of cost, you can buy an ionization type alarm for $8; a photoelectric or dual alarm for $25; and a photoelectric plus carbon monoxide alarm for $40.

If your home does not have any smoke detectors, I recommend you make it a priority to purchase the photoelectric type ($25) and get the wireless models that communicate with each other. Install smoke alarms on every floor of your home in hallways outside bedrooms, inside bedrooms, in other main hallways, and just outside the kitchen area.

Consider installing the dual photoelectric – carbon monoxide units. If someone leaves a car running in the garage or your fireplace is producing carbon monoxide you will have enough warning to leave the house.

Escaping a house fire is all about time. The faster the warning, the sooner you can escape. Photoelectric alarms provide a fire warning in half the time the ionization units do. This provides precious time for you and your loved ones to get to safety.

Choosing Smoke Detectors: What You Need to Know

NBC Nightly News ran a story about home fire detectors last week. The story was particularly interesting because of a recent study revealing that 90% of homes have ionization technology smoke detectors rather than the more expensive photoelectric technology detectors. In the study, it took twice as long for the ionization models to sound the alarm compared to the photoelectric units. This represented about 20 minutes – a long time. Time is one thing that we do not have enough of in a fire where seconds count. You may already know that about 70% of the deaths in residential fires are a result of smoke inhalation. We also know that detectors are essential in warning us of fire. In light of this dramatic news story, what should you do?

First, some statistics. According to the National Fire Protection Association (NFPA), non-working detectors exist in 20% of the homes with alarms installed. As a home inspector, I can tell you from experience that this statistic is on the low side. Every third home I inspected had non-working fire detectors. Over half of the nation's residential fire deaths happen in homes without any detectors at all, and the overall risk of dying in a home without working smoke detectors is twice as high as the risk in homes with working detectors.

What types of smoke detectors are available and what type should you have in your home? Ionization smoke detectors are designed to sense fast flaming and fast moving fires. These particles may be invisible to the eye. This type of detector will go off quickly in the high heat of a flash fire and a fire that has relatively little smoke associated with it. A photoelectric smoke detector detects visible particles, as you would have in a fire that has barely begun, but is producing smoke. An example of this would be a cigarette falling into upholstery and smoldering.

What to do? First, if you already have detectors installed, take a look at them. Write down the identifying information and look up the type (ionization or photoelectric, or both) on the internet or take the information to your favorite home improvement store so that they can identify it for you. If you discover that your detectors are only the ionization type, I recommend that you purchase additional detectors of the photoelectric type to supplement your current detectors. It is also possible to purchase detectors that have both technologies in one unit. Follow the directions for usage, testing, and maintenance. This includes periodic testing (test button), replacing batteries annually, and keeping vents clear. You can also purchase detectors that are completely battery operated and communicate wirelessly with all the others so that when one goes off, they all go off.

If your home does not have any smoke detectors, I recommend you make it a priority to purchase dual technology models (start at $18) and get the wireless inter-communication models if you have a little more money to spend. Install smoke alarms on every floor of your home in hallways outside bedrooms, inside bedrooms, in other main hallways, and just outside the kitchen area. Taking these simple steps now will afford you peace of mind in light of the recent news and provide an extra margin of safety for you and your loved ones.

Fire Safety Quiz

When I inspected homes, I paid a lot of attention to safety items. It was one thing to find a window that would not shut, and another to find half of the smoke detectors not working.

A few weeks ago, Felicia Mull of Clay County Fire and Rescue reminded us that fires are more common in the fall as we start up the wood stove and fireplace. Are you ready? Take this quiz and see! Pick the best answer.

1. Before you throw wood in and light the fire in your wood burning fireplace you should: A. Arrange the furniture around the fireplace so everyone can see it. B. Make sure the logs are less than 3 inches in diameter. C. Start the fire and then leave it unattended while it warms up. D. Inspect the chimney for soot that could catch fire.
2. Modern smoke detectors are self-testing and you don't need to check them. True or False?
3. What are the two technologies that you should look for in a modern smoke detector? A. Breathalyzer and Misting. B. Particulate and Anti-ionizing. C. Photo Effect and Humidifying. D. Ionizing and Photoelectric.
4. Your wood fireplace can also be used to: A. Send smoke signals. B. Burn your trash so you don't have to take it to the dump. C. Heat the entire house while you are at work. D. A, B. and C. are wrong and not safe.

5. Something that you and your family should practice several times a year is: A. Have a contest to see who can make the biggest fire. B. Put smoke chips in the fire and wave the smoke into the room as an air fragrance. C. Disable the Carbon Monoxide detectors in the winter. D. Practice a fire evacuation drill twice a year.

ANSWERS:

1. D. Inspect the chimney for soot that could catch fire. Thousands of fires each year are caused when soot catches fire in the chimney itself. These fires are very difficult to put out. Have your chimney inspected and cleaned on a regular basis.
2. False. You should push the "TEST" button on your detectors right now and on a schedule. Put this minor chore on your checklist or calendar. On inspections I found alarms not working or missing batteries 60% of the time.
3. D. Ionizing and Photoelectric. Smoke alarms that use ionization technology are great at detecting a fast, flaming fire such as burning paper, but poor at detecting a smoldering fire, as in a couch or mattress. Photoelectric smoke alarms are good at detecting a smoldering fire (with smoke). I recommend you get units that use both technologies. If you are uncertain what you have, your chimney sweep or handy person will be able to check them.
4. A, B. and C. are wrong and not safe. A lot of smoke from your fire will leave some to come out into your room, polluting the air. Burning your trash will also allow toxic fumes into your room and could get out of control. Never leave a fire unattended.
5. A, B. and C. are wrong and not safe. D. is the correct answer. Put a plan together with your family that includes materials, such as a rope ladder if you have a second story in your home. Make it fun with rewards for the fastest times.

How did you do? Be prepared and do not allow a fire to take you by surprise. If you do discover a fire, call 911 and put your emergency evacuation plan into execution.

Six Safety Items No Home Should be Without

The hustle and bustle of Christmas is over, and as we settle down in the easy chair for another round of football playoffs, we look around at the post-holiday mess. The guests have done our dishes but put them back in different drawers and cupboards. The boxes for the decorations have gone missing. Gifts, tools, cards, keys, books, and electronics lie on tables and chairs. Soon it will be time to go back to work and school. When will we have time to deal with all of this? Don't think about it. Watch football.

But I am thinking about it, as I take another look around. What if we had a sudden power outage, or a fire broke out? It would be great to be organized ahead of time for these uncertain events. Remember Mr. Murphy – what can go wrong will go wrong – so the corollary is, if you are prepared for something to go wrong, it won't go wrong. Unless you live in Murphy. Murphy times Murphy = Murphy Squared, which means that under the very best of circumstances several unrelated things will go wrong at the same time and appear related.

So, after the games make one of your New Year resolutions to outfit your house with the following items to stave off Mr. Murphy.

1. Fire Extinguisher. Keep a fire extinguisher handy in the kitchen, in the garage or carport, and in the second story area if your home has more than one level. I like the Kidde FA110 Multi-Purpose Fire Extinguisher 1A10BC, available locally or online for less than $30. It is designed to fight basic fires,

including those involving fabrics, plastics, wood, flammable liquids, and electrical equipment. It has a charge gauge and instructions right on the cylinder.

2. Flashlights under every sink and by every bed. Modern flashlights, especially LED versions, stay powered longer. As your old metal housed "D" eating flashlights die off, replace them with the new versions.
3. Smoke hood in every bedside table. Remember Murphy – if you have it, you won't need it – but if you do wake up to a fire then toxic smoke will be your biggest enemy. A smoke escape hood can be purchased for less than $15 and can save a life. Check for them locally or go to Amazon.com.
4. Emergency ladder. If you have a second story in your home can people get out and to the ground safely in the event of a fire? If not, purchase a simple emergency ladder and keep it handy.
5. Whistles. Keep some whistles handy on each floor and tell members of the family to blow them if they spot something going wrong fast, such as fire breaking out or a bear breaking in (don't use the whistles for things like the internet went out or the Steelers lost). Decide what the triggers are so everyone knows. If you're by yourself they are handy for getting attention from emergency responders.
6. First Aid Kit. It goes without saying you should have a fully stocked first aid kit in the kitchen at a minimum. Keep one for emergencies, and one for daily use. That way you'll always have one that doesn't have stuff missing.

This list should prompt you to think of other important items. Write them down. Outfitting your home with the minimum items listed above should cost less than $200.

Tips for Safe Holiday Decorating: Ladders

For a holiday time of year, it certainly seems like there is a lot of work going on around the house. When you add decorating, hanging wreaths, putting up lights, assembling or getting the tree, card writing, gift buying, gift wrapping, cooking, and entertaining . . . it can be overwhelming. We really ought to have 2 weeks off before Christmas and 2 weeks after Christmas just to get things set up and then get them, and ourselves, back to normal.

One of the simpler things I enjoy is putting lights on some of the evergreen trees outside. Sparkling, colorful and bright, the decorations bring excitement and cheer to a dark winter landscape. Whatever your fancy, many of the decorating projects such as this one involve the use of ladders.

Have you ever had a scare while using a ladder? Worse, have you ever tripped or fallen while using a ladder? If not, you're lucky. More than 40% of the population has slipped or fallen while using a ladder or stepstool. Even the best preparation can lead to injury if we're not paying attention. Here are some tips for staying safe both now and all year round.

Look the ladder over before you place it. This is particularly true of wood ladders, extension ladders, and step ladders. Check rungs for condition and make sure they are tight. Look at moving parts on extension and step ladders along with the extending cord while the ladder is on the ground.

Assemble all the things you need before you climb the ladder. Have you ever gotten up to the eave or on to the roof and said, "Shoot, I forgot the staple gun!" Going back for something makes you hurry and is distracting. Slow down and be careful.

Set your ladder up at the right angle to the house. Put your toes against the ladder's feet. Stand straight up and extend your arms. The palms of your hands should just reach the ladder's rung. Next, check 2 things: will the feet slip on the surface (such as loose soil or pebbles) and is the surface level? A slope of only an inch will magnify the error considerably by the time you reach the top of the ladder.

The ladder should extend at least 3 rungs over the roof line if you are getting on to the roof. This adds stability and helps prevent the ladder from shifting. One practice that I like is to take a couple of bungee cords with me and secure the ladder to the trim or gutter brackets. Don't count on this to provide strength to the positioning, but it will add an extra bit of stability.

Stay centered on the ladder so that your balance point is straight down. Yes, this means you can't lean way out over the side. You will need to reposition the ladder more often, but this practice is a lot safer.

Women have one quarter of the accidents on ladders that men do. I suppose this either means we are extra careful or we are really good at talking the guys into climbing the ladders!

Kitchen Safety

The kitchen is a warm and inviting room to be in during the holidays. The smell of cookies and cakes baking, and the warmth of the oven emanate out to other parts of the house and draw us to this area of pleasures. The kitchen tends to attract family, friends, and guests. Sometimes there is no reason to leave the kitchen, as we sit or stand at the counter or kitchen table enjoying the activities.

Not to spoil the mood, but the kitchen can also be a place of danger. According to the Consumer Product Safety Commission, the kitchen is a top area for injuries. It's easy to understand why; from knives, to slicers, to hot burners and ovens, the opportunity for injury is in every corner. Knife lacerations alone accounted for over 350,000 injuries in 2012.

Here are some safety tips to keep that bright, warm mood in the kitchen year-round.

- Fire extinguisher. Do you have a fire extinguisher in your kitchen? If not, purchase a 5B:C rated extinguisher to keep handy in a cabinet corner or out of the way on a wall. If you never need it, great. You can also use baking soda to dampen a grease fire. Don't leave the kitchen when you're cooking on a stovetop, you could get distracted and not return in time to catch a problem.
- Sharp knives cause fewer injuries. Counterintuitive, but true. Dull knives require more hand pressure and slip more easily than sharp ones. You don't need to purchase a fancy sharpener; a small handheld sharpener will do wonders.
- Anti-tip brackets on your range. "Anti-what?" you say. Anti-tip brackets are actually required, but rarely used, on the back of your range to keep the range from tipping forward when the door is open. This can happen when a child climbs on the open door, or if a heavy pot or pan (think turkey) is pulled out and the weight shifts forward. To find out if your range has this safety feature installed, simply pull on the range from the back forward. It should be impossible to tip forward. If you find that the unit does tip up, run down to your favorite hardware store and ask for this bracket. They will know what it is, and be able to help you locate someone to install it if you're not comfortable doing it yourself.
- Microwave Ovens. You already know that you're not supposed to reach in and grab the food package right after the buzzer goes off. I don't know about you, but I do this, and have paid the price. Microwaves are wonderful time savers, and since they heat the package from the inside out, most of the packages are not hot on the edges. However, the food itself IS hot, so you should be careful. Use gloves or mitts if you've got something you know is hot to protect yourself.

- Slicers, Choppers, and Blenders. Resist the temptation to put your hand into the works when these units are plugged in! As simple as this sounds, thousands of injuries are caused this way every year. When cleaning blades, put some detergent into the device and operate it, then rinse.

Enjoy your kitchen and stay safe throughout the year.

Protecting Yourself and Your Home From Lightning

When I was very young I remember standing at a large plate glass window in the living room and seeing a flash of lightning. It was dramatic and frightening to my five-year-old senses, and the thunderclap that followed a moment later caused me to run from the living room to the kitchen, where my mother was standing, looking out at the rain pelting the kitchen window.

"I love a good storm," she said to me. "It's fun!"

I shook my head in bewilderment.

"As long as we are inside, safe, we are fine," she reassured me.

I was not entirely sure that she was correct, but luckily I grew into an adult enjoying the fury of the outside elements, realizing I had no control over them, and believing that I was safe inside.

Are we safe? For the most part, yes. Statistically your chances of being hit by lightning while inside your home are less than 1 in 300,000. This is pretty low. But it does happen. Every year several hundred homes get a direct lightning strike, and if that happens when you are in the shower or talking on a corded phone, the charge can injure or kill you. In the U.S. about 50 people die from a lightning strike every year, but most of these (45 of the 50) are outdoors.

How to protect yourself inside your home.

- Don't stand near open doors or windows. When lightning strikes, it spreads out from the strike.
- Avoid contact with water. This means not showering, taking a bath, washing dishes, or doing laundry. Electricity runs along wiring and metal pipes, and water is a conductor.
- Don't talk on a corded telephone during a storm. Wires can carry electricity straight to you. About 5% of all fatalities from lightning strikes are people talking on a corded telephone. Using cordless phones and cell phones are safe.
- Consider not using the computer or TV while a thunderstorm is raging. A lightning strike to your home or even nearby can cause a surge that arrives via your electronics.

How to protect your home:

- Lightning's threat to your home is in the form of fire. According to the U.S. Fire Administration, at least 8000 homes catch fire as a result of a lightning strike. If you observe smoke or any other sights and smells that indicate fire, you should call 911, get out and drive far enough away to observe. Cars and trucks are quite safe compared to being in a burning building or out in the open during a storm.
- Electronics. Remember that the best of surge protective devices will not stand a chance if your home is struck by lightning. Surge protectors are just that – they protect your electronics from the ups and down of the electrical supply. I would still recommend surge protectors, and a good UPS (uninterruptable power supply) for your computer. If you are home during a bad storm, consider unplugging major electronics, like big screen TVs and computers. Some electric companies (like BRMEMC) offer protection through insurance and suppressor devices which they install and monitor for a monthly fee.

- Lightning Rods. A "lightning rod" or lightning protection system, will not keep lightning from striking your home. But if lightning does strike your home, a professionally installed system may keep your home from catching fire by directing lightning to the ground around the home. These systems are usually installed when the home is being built, since a ground grid that attaches to cables and rods at the top of the home are buried around the perimeter and run inside walls. These systems can be installed for an existing home, usually for several thousand dollars. If you seem to get a lot of strikes near your home, you may want to consider this protection.

Now that you have all the facts, enjoy the fury of the next big storm. Too bad we can't explain this to our dogs and cats.

Fall Chores and Ladders

A pleasant cool breeze this past weekend reminded me that fall is around the corner. On my fall "to do" list I have "deck stain" once again because I never completely finished the job from last summer. Temperatures for stain application are pretty forgiving, but once the temperatures really drop, the chore has to be shelved. Oh good! Put the ladders away!

Many chores around the home involve ladders. Even if your toolbox doesn't have ALL the tools you really want, I expect you do have plenty of ladders. Rushing to complete fall chores combined with an assortment of ladders can be dangerous.

According to the World Health Organization, the United States leads the world in ladder caused deaths. Each year there are more than 160,000 emergency room visits and more than 300 deaths that are caused by falls from ladders.

Most ladder deaths are from falls of 10 feet or less.

Injuries from ladder use have tripled over the last 10 years.

Here are some ladder safety tips that will help you avoid spending 12 or more weeks in a cast, or worse. Resist the temptation to say, "I already know this." If even one of these tips makes you pause to think, it will be worth it.

- Look the ladder over before you place it. This is particularly true of wood ladders, extension ladders, and step ladders. Make sure everything works and that rungs are not loose, rotten, or bent.
- Assemble all the things you need before you climb the ladder. Have you ever gotten up to the eave or on to the roof and said, "Shoot, I forgot the staple gun!" Going back for something makes you hurry and is distracting. Rushing and not paying attention are big causes for accidents.
- Set your ladder up at a 75-degree angle to the house wall and look carefully at where the feet are. You want to check 2 things: will the feet slip on the surface (such as loose soil or pebbles) and is the surface level? A slope of only an inch will magnify by the time you reach the gutters.
- The ladder should extend at least 3 rungs over the roof line if you are getting on to the roof. This adds stability and helps prevent the ladder from shifting. One practice that I like is to take a couple of bungee cords with me and secure the ladder to the gutter brackets. These are not strong, but do add stability.
- Stay centered on the ladder so that your balance point is straight down. Yes, this means you can't lean way out over the side! And yes, you will have to move the ladder more often. Would you rather spend a little more time moving the ladder or spend time in the emergency room?

Finally, if you are over 50 listen up. We "older folks" don't have the same sense of balance that we had when we were younger, and our reflexes are a little slower. This means taking even greater care using ladders. Use that wisdom that develops with age and hire the teenager down the street who needs the job!

Ladder Safety

Is it time to put up the Christmas decorations? You bet! Let's get out the ladders and . . .wait! Before you jump on those ladders dragging strings of lights, wreaths, and extension cords behind you, let's talk a little bit about the ladders.

You may never have had a close call on a ladder. Good - you're probably doing something right, or you are lucky. There are plenty of us, including myself, who thought we were doing everything correctly and still found ourselves either having that heart in the throat scare or found ourselves falling. In fact, 70% of the home inspectors and contractors I've talked to about ladders have either had a bad scare or have fell. These are people who are trained in safe ladder use and generally have well maintained equipment. By the way, women have a quarter of the accidents men have on ladders. I think this is because we have a more developed sense of self preservation and are not risk takers to the degree men are. Either that, or we're just better at talking men into doing these types of chores!

Now that I have your attention, here are some ladder safety tips that will help you avoid spending 12 or more weeks in a cast, or worse. Resist the temptation to say, "I already know this." If even one of these tips makes you pause to think, it will have been worth it.

Look the ladder over before you place it. This is particularly true of wood ladders, extension ladders, and step ladders. Make sure the integrity is 100% and that rungs are not loose, rotten, or bent.

Assemble all the things you need before you climb the ladder. Have you ever gotten up to the eave or on to the roof and said, "Shoot, I forgot the staple gun!" Going back for something makes you hurry and is distracting. Rushing and not paying attention are big causes for accidents.

Set your ladder up at a 75-degree angle to the house wall and look carefully at where the feet are. You want to check 2 things: will the feet slip on the surface (such as loose soil or pebbles) and is the surface level? A slope of only an inch will magnify by the time you reach the gutters.

The ladder should extend at least 3 rungs over the roof line if you are getting on to the roof. This adds stability and helps prevent the ladder from shifting. One practice that I like is to take a couple of bungee cords with me and secure the ladder to the gutter brackets. These are not strong, but do add stability.

Stay centered on the ladder so that your balance point is straight down. Yes, this means you can't lean way out over the side! And yes, you will have to move the ladder more often. Would you rather spend a little more time moving the ladder or spend a lot more time in the emergency room?

Finally, if you are over 40 take heed. We "older folks" don't have the same sense of balance that we had when we were younger. This means taking even greater care using ladders. Use that wisdom that develops with age and hire the teenager down the street who needs the Christmas money.

Attached Garages: Luxury or Danger?

My first home was a small ranch house, with a car port. I remember thinking that this was absolutely the cat's meow – a roof over my car! In fact, the roof line came out far enough for me to actually walk from the front door of the house to the car in a pouring rain without getting wet. I'd arrived.

Since that time, I've become unaccountably spoiled and the next house I build will now have a basement garage! This is wonderful – in the winter I won't even have to go outside. This car-home arrangement is called "an attached garage." There's no doubt that this is very convenient, even luxurious.

But hold on. Danger lurks. Attached garages can produce a variety of problems that detached garages do not have. We have heard the story about someone driving in to the garage and forgetting to turn off the vehicle, causing death to home occupants from carbon monoxide poisoning. More often, attached garages provide a source of pollution to the house that can cause unexplained health complaints. Here are the two biggest attached garage dangers and what you can do about them.

Carbon Monoxide. "CO" is an odorless, colorless gas created when fossil fuels (gasoline, etc.) are burned. Running the car in the garage builds up CO concentrations fast, even with the door open. The gas can then enter the home environment if the seal between the garage and the home is not tight. What to do:

- Take a good look inside your garage. You should see firewall rated walls and doors and absolutely no light around the door to the house. If you have any ductwork in the garage, you should see mastic compound on joints. If you have an older home and are worried about airtightness, call a home inspector or energy advisor. They will be able to identify leaks and recommend solutions.
- Once you get in the car, open the garage door before turning on the ignition. Move out quickly, giving the garage a few minutes to ventilate before closing the door. When returning, don't let the car keep running, and leave the door open for a minute or two to ventilate.
- Consider installing an automatic vent fan in your garage to ventilate the area after the door closes. Some vents can come on with detected motion, and turn off automatically. A garage vent will reduce the tendency of your house to create negative pressure, pulling in garage air.

Fire. An attached garage fire can be devastating. Once a fire begins in a garage, it literally explodes into the home if not contained. Statistics show that the garage is a prime area for fire starts in residential homes. What to do:

- What are you storing? If you see paint, gasoline, oil, adhesives, and thinners, either move these to an outside shed, or buy a good flammables cabinet for storage. Be disciplined about putting them in the cabinet when not in use.
- Electrical cords. In garage workshops it seems as if there are never enough outlets. If you have extension cords plugged into strip outlets like Chevy Chase in Christmas Vacation, all it takes is a spark to ignite fumes or chemicals.
- Install both fire detectors AND carbon monoxide detectors in your home to give you early warning of a problem. Follow the installation directions and test regularly.

We take our garages for granted – they are convenient – but don't be caught off guard. Inspect your garage for hidden dangers and stay safe.

Should You Be Concerned About Product Recalls?

Should you be concerned about product recalls? Do you know what products in your home have been recalled by the manufacturer? Your response might be, "I have enough other things to worry about besides if the legs on my side table are going to fall off." Indeed, our world is so full of information now, having to consider even one more thing can be overwhelming. However, I am going to argue that keeping track of some of these things – especially children's products, computers, and appliances – might be a good idea.

A few examples of cases where it would have been good to know the dangers ahead of time include combustible computer batteries (Hewlett Packard, Sony), refrigerator doors falling off (Viking), bursting fitness balls (Bally, Everlast), parts falling off weedwhackers (Black & Decker), bursting flushing systems (Flushmate), and toxic Aqua Dots craft kits (Spin Master).

Although your favorite thing is not sending in the little registration cards that come with products, if it saves you or a family member from even a minor injury, it will be worth it. Here's a system to stay on top of recalls.

First, do send in the registration cards. If the product is a big purchase, like a computer, an appliance, or a child's product, don't just send it in, make a copy of it and place it in a folder called "Product Registration." Manufacturers are required to notify you of recalls or safety defects, so it might be a good idea to include them on your mailing list if you move. This way a manufacturer can also notify you of any new maintenance advice to improve the longevity of your product.

Second, check the products that you already own by going to www.recalls.gov. If you are not internet savvy, ask the teenager down the street to help you. The manufacturer information is included on this site so you can write to them if you are having a problem with the product.

Third, check products out before you buy them. Go to www.SaferProducts.gov to see if the product is listed. The site is easy to navigate and full of information. Even though it is illegal to re-sell a product that has been recalled, 90% of flea markets, eBay sellers, and garage sale sellers don't have a clue that a particular item has been recalled, and I wouldn't expect them to.

As a home inspector, I always check recall status on major systems and appliances for the buyer by going to the recalls website I mentioned above. This can be a major safety issue that unwary home buyers can stumble into. One example is an older model furnace that would overheat, causing the heat exchanger to crack. The manufacturer recalled the units, but many homeowners did not find out there was a problem until fires started in the furnace.

Following the advice here is a little more work but can be a positive safety net for your family. Keep this notebook with your maintenance records and schedules and you'll know you did everything possible to stay safe in this complicated world.

12

Basement and Crawlspace Problems

Everything You Didn't Want to Know About Crawlspaces

The lovely crawlspace. The name itself sounds creepy. The space where things crawl? Indeed, as a home inspector I got used to seeing crawling things, both dead and alive, in the spaces underneath the homes I inspected. Lucky for you - I'll save these stories for another column.

If you have a wet or damp basement, the problem makes itself obvious quickly. If you have a wet or damp crawlspace, you might never know about it until your home starts smelling musty or the air seems damp and cold, or you notice mold starting to grow on baseboards.

Problem crawlspaces can exist for a long time without appearing to cause problems, but eventually those problems will surface and reduce the value of your home. If left uncorrected, the humidity in a crawlspace can cause air and heating components to rust, wood flooring structure to deteriorate and buckle, bolts and metals to corrode, provide an invitation to termites, and encourage the growth of mildew, fungus, and molds.

What should you do?

Simple. Go look.

Winter is a good time for this inspection. The creepy crawlies in your crawlspace are either hibernating or sluggish, so when you open the door to enter, you shouldn't see anything moving. If you get cold feet reading this, then you might want to ask someone else to do this inspection. Here is what they should look for:

- Is there any standing water? Is anything dripping from understructure or piping?
- Is there any discoloration on the wood members or on the insulation under the floor?
- Is insulation falling down?
- Is there bare dirt on the floor of the space?

If you answered yes to any of these questions, you may have a problem. The first three issues mean that there is too much humidity in the space. High humidity greatly accelerates rot, rust, and the formation of a variety of molds. The fourth condition, bare earth, aggravates the first three.

You can call a contractor to fix these problems, or you can try fixing them yourself. If you try solving these yourself first, you should limit what you do to three things. The first is making sure that the grading around your home is leading water away, not towards, the crawlspace. The second is to install 6 Mil plastic over the dirt floor, leaving a small margin between it and the walls. If these two solutions don't solve the problem, then try installing several small fans in or near the vent windows to get the air circulating. You can even install a de-humidifier in your crawlspace.

If these steps do not help, then it is time to call a professional. Mold specialists in our area can determine what needs to be done and also perform the work. Make sure the person you hire is certified to work with mold. These specialists can even "encapsulate," or seal, your crawlspace.

Taking care of your crawlspace will pay off in longer home component life, less chance of a termite infestation, and healthier living conditions inside the home.

Crawlspace vents – should they be closed in winter?

A Progress reader writes, "We have sealed up the drafty areas in the house and it feels a lot cozier. Our floors are cold though. We have a dirt crawlspace and we don't know whether to close the vents or not in winter. What do you advise? We don't have any insulation under the floor, and the space is just bare dirt."

I get this question all the time on inspections and it's a good one. There are several things you can do, depending on how much or how little money you want to spend.

At a minimum, I would cover the dirt in the crawlspace with 12 mil plastic. This will stop dampness from rising out of the ground and penetrating your floor structure, and it will keep the space warmer. You can get this "vapor barrier" at local home and hardware stores or online at www.crawlspace.net. If your crawlspace is fairly dry and humidity is not creating mildew on your floor under-structure, then I'd close the crawlspace vents to help retain heat.

If you can spend the money, I would then add insulation to your under-floor area. This one step will make a large difference in how well your home retains heat (and reduce energy costs).

If you want to go one step further, close off your crawlspace vents completely and add a dehumidifier to the space. Technically, crawlspaces and attics are supposed to have a certain amount of ventilation, and building inspectors inspect to these standards on new homes. The "green" building movement has produced some variations, however, that contribute to energy savings but have to meet the inspection requirements for ventilation – hence the addition of standalone recirculating systems. These systems keep the structure clean and dry and mold free.

On some inspections, I have crawled through spaces that were so damp that water was dripping off floor joists, floor boards were rotten all the way through, mold was growing on the wood structure and the walls, and I was crawling through puddles (in a waterproof suit!) throughout the space. The owners complained about cold and damp floors and did not realize there was a problem until a member of the family stepped on a rotten section in a closet and ended up in the crawlspace!

If you have these kinds of problems, I would definitely hire a qualified contractor to review your situation and advise on next steps.

If your crawlspace is humid and damp now, I recommend solving that problem first, before you close the vents. Adding the plastic to the dirt may be the complete solution, but try it and see.

Leaving the vents open in winter will not hurt anything and will probably help lower the humidity by introducing ventilation. The downside is a colder crawlspace, and potentially frozen pipes if your pipes are not currently insulated in that area. Ideally, you can add insulation under the floor AND the vapor barrier, and close your vents in the winter if you find that the humidity is 60% or less. This will keep your floors warm, and your pipes safe.

Is Your Basement Damp or Wet?

Summer in the mountains - a lovely time to see the rain fall and nourish our garden plantings. Thunderstorms build in the west and create amazing cloud formations. The bushes we planted last year are taller, fuller. The foliage is thick and green. Water from the fresh rain is trickling down the slopes past our home. All is well. But is it?

Do you have a crawlspace or basement? Here in the mountains it is common to have our homes on a slope and our crawlspace or basement is half in, half out, of the mountainside. Or we might be in the floodplain next to a gurgling stream with our home foundation set up a few feet from the meadow. Unless you have a basement and detect a musty odor . . . the telltale sign of high humidity - you may not pay any attention at all to your crawlspace until you notice the advanced signs of the number one enemy of your home - moisture - making its way into your home's rooms as you notice wet carpeting, stale odors, damp baseboards, or damp foundation walls.

As a home inspector, I encounter damp basements and crawlspaces about a third of the time. If left uncorrected, the humidity in these spaces can cause HVAC components to rust, wood flooring structure to deteriorate, bolts and metals to corrode, cause termites to make these spaces their new home, and encourage the growth of mildew, fungus, and mold.

Here's a tip that will help keep you and your home healthier. Along with your spring planting and cleaning tasks, go into your basement or your crawlspace (ok, get someone else to go into your crawlspace!) and take a moment to observe everything you see. Do you smell moisture or mildew? Is the humidity high? Do you see any condensation on the walls or the HVAC ductwork? Is water dripping anywhere?

Taking the time to inspect these areas periodically will pay off in fresher air, less pest intrusion, and longer system component life. All it takes is a clogged gutter or drain, or a wet spring as we had here in Clay County, to generate damaging moisture intrusion in your home.

If you discover a wet basement or crawlspace, there are some simple things you can do yourself.

First, check the ventilation ducts (basement) or vents (crawlspace). Make sure these vents are not closed or sealed during the summer, as ventilation is one of the keys to lowering humidly.

You can try putting a box fan in your crawlspace to dry it out, or consider vent fans. If you have a crawlspace and you see condensation, make sure you have a plastic vapor retarder installed over the dirt.

Bare dirt generates gallons of moisture and you don't want this moisture hanging around under your house.

Next, check the water runoff around your home. Are gutters working and are they depositing water AWAY from the foundation? Is there a gap between the soil around your home and the siding? If there isn't, then the soil is retaining water against your foundation and adding moisture to the space behind it. Pull the soil back from your home so that you have at least six to eight inches between the ground and the siding. Are your beautiful bushes growing right up the sides of your house? Vegetation can hold moisture and deposit it on your siding and next to your foundation. Trim the bushes back so that you have at least 12 inches of horizontal clearance between your plants and your foundation wall or siding. This will also help prevent termites from deciding that your siding will be their next food stop.

Now wait for a heavy rain. Go outside (make sure you're not in a lightning storm) and look at where the water is going. Is it running against your home? You may need some revised grading, drainage, or landscaping. Are the gutters working or is water pouring off the roofline? After the hard rain you will know where to make the improvements to keep water away from your home.

Lastly, you may need to run a de-humidifier in your basement to dry it out. Get the humidity down to 60% or less. You can purchase an inexpensive hygrometer in most general stores to find out what the humidity level really is and most de-humidifiers have automatic settings so you can set it and forget it. This is an especially good idea if you are not in your home all the time (vacation home). Be sure to follow the directions for running the condensate (water) line to an appropriate drain.

Don't hesitate to call in a professional contractor if these steps do not solve your problem. There are easy corrections that can be made to dry these areas out and head off future problems. Keeping your home dry will extend its life, your comfort, and protect your investment.

Solutions for Damp Basements

The words "damp" and "basement" seem to enjoy each other's company, even though we are not happy about the friendship. There are many reasons why it is common to have damp or even wet conditions in our soil surrounded rooms, and some of these conditions are difficult to solve.

If humidity is high enough, mold, mildew, and bacteria grow quickly. Dust mites thrive in high humidity and high humidity can worsen asthma and allergies. Mold will ruin carpet, wood, insulation, and fabrics.

How do you know if you have dangerously high moisture levels in your home, and what can you do?

First, find out what the humidity actually is in any area where you experience damp conditions. Place a hygrometer (a device that measures how much water is in the air), available at any hardware store, in the room and leave it there for several hours. Take a reading and write it down. Wait 8 hours and record it again. After you've taken readings for several days you should have an accurate average reading. If your average reading is anything over 60%, you may be developing a problem. The rule of thumb for optimum conditions is 35-55% RH (relative humidity). In the summer, 40-50% is ideal.

The following solutions cover 80% of the causes for wet or damp basements. If you think your problem falls outside the following causes and solutions, you should call a professional contractor for advice and repairs.

- Gutters and drainage. During a one inch rain over 600 gallons of water will fall on a 1,000 square foot roof. Make sure this water is adequately carried off and doesn't land on the ground right next to your walls. Downspouts should extend out at least 5 feet from the house.
- Your home is brand new. New homes will hold extra moisture in the framing, the drywall, and the concrete for months

after construction. My advice is to first make sure your new heating and cooling system is balanced and sized properly for the zones, and secondly to purchase a dehumidifier. Run the dehumidifier until the average humidity lowers to 50 - 55%.

- Grading or landscaping was not done correctly and/or the soil has settled. Measure the grade fall-off outside your home. The landscape should slope down and away from your home for at least 10 feet (six inches drop over 10 feet is a minimum). If it does not, then moisture is being trapped against the basement walls. To solve this problem, you should call a grading contractor who will adjust the fall-off so that water is led away from the house.
- Water or moist air leaking through the walls or floor. This is a much more serious problem, but is not uncommon. Any cracks in the floor or walls will admit moisture. You may be able to seal these yourself. If there is a lot of cracking or obvious gaps, you should call a professional for the repairs.

These solutions cover the majority of problems that I've seen. A dehumidifier may be required in your basement even after you do everything right. Very efficient and quiet units are now on the market and cost a few hundred dollars. Go to Consumer Reports for information on the best value for your room size and wallet.

13

What to do about Mold and Humidity

Your Healthy Home: Mold in the Attic?

A progress reader sent me an email last week with a question about mold in her attic. Cindy writes: "I just discovered very damp boards in my attic where I store Christmas items and files. I've never seen this before and my roof is only 3 years old. Is this a result of the exceptionally cold winter, and what can I do about it?"

Typically we think of mold as growing in the basement, where conditions can be damp, and not in the attic; but mold can and does grow anywhere we have a common element: moisture and the resulting high humidity. While in this short column I cannot go into great detail, I will try to point you in the right direction.

Mold comes in 1.5 million – yes, that's right – 1.5 MILLION species. That's a lot of mold! Mold is critical for the biological breakdown cycles of the earth, such as on the floor of the forest, where leaves and plant debris turn into mulch and eventually fresh earth through this "moldy" breakdown process. It's important to note that not all molds are harmful to humans. We often associate mold with illness or allergy, but there are a limited number of species that are harmful to us. Also, each of us has a different level of sensitivity to molds. Hundreds of tiny mold spores float around in our air – both inside our house and out

– and do not harm us until the point where the concentration grows dramatically – and it is moisture that triggers mold growth.

In the attic, there are three ways to get mildew or mold: inadequate ventilation, a roof leak, or some other source of moisture. First, check what type of ventilation system you have. Ridge vents at the peak of your roof? Soffit vents at the overhangs of your roof? A powered fan? Check these to make sure they are working. Is insulation blocking your soffit vents? Second, do you have a roof leak? Even if it's a new roof, you can have leaks at plumbing stacks (pipes) that were not gasketed properly. Now check for a source of moisture. Are your attic stairs letting warm air from your home into the attic? If it is cold outside (cold in your attic), this can form condensation, or water droplets to form and drip off various surfaces in the attic. Inspect the seal at hatches to the attic also. One big contributor to attic moisture that I see 40% of the time in the homes I inspect is where the builder did not run the bath fans OUTSIDE as most building codes dictate, but into the attic. You then have a situation where you are venting hot moist air from the bathroom directly into the attic. The result is high attic humidity, condensation in cold weather, and possible mold conditions.

What can you do if you think you have mold in your attic?

First, buy a hygrometer. This measures the relative humidity in the air. You can buy one at your favorite home supply store or hardware store. Take a variety of readings in your attic in different spots. If you are seeing over 50 - 55% relative humidity consistently, you have a problem. What you want to see, ideally, is 35 – 45% R.H. If it is high, you know you need to find and eliminate the moisture problem, then deal with the mold problem. And it may not be as bad as you think: sometimes the material the mold is on can be dried out and just left alone.

If you think your problem is complicated, extensive, or serious, you should consult a professional IAQ (Indoor Air Quality) certified tester. This person can tell you what you have and how to get rid of it, and these contractors usually also perform the services to remove it if that is necessary. Look in your local business listings, check with your Chamber of Commerce, or look on the internet for local companies that are certified to perform these services.

Lastly, do get advice or do some research before trying to clean up mold. Mildew on a tile wall is straightforward and household chemicals will take care of it, but mold on other surfaces, such as wood or insulation will be made worse by using cleaning products such as bleach or disinfectants. A good independent web site to find out more is: www.inspectapedia.com.

Your Healthy Home: Mold in the Basement part 1

Last week we found out that attics – the places that we thought were always dry – the place where we put Aunt Minnie's 75-year-old portrait - can actually be wet and grow mildew and mold if not corrected. However, this is not the norm. What is far more common is finding that you have mold in the basement, not the attic. The varieties of wet basements are so numerous that we may not be able to cover them all, but we'll hit the important things to know and do to solve the problem if you have it.

As we said in last week's column, the development of mold is a natural event, but we don't want it in our homes! The spores are in the air and land everywhere in our homes, but without the recipe for growth – moisture and a porous surface – they will not activate.

Understanding basic physics will help us understand why mold gets into our basement. Rain falls; streams run downhill; vegetation grows up and in, and Murphy's Law states that even though you just cleaned your gutters, you will have overlooked one section which will clog and water will overflow from your roof to the front of the house and run into the basement.

If you have mold in your basement, start solving the problem by looking outside first. Go outside and walk away from your home about 100 feet. This is the first thing home inspectors do when they inspect – stand there in the front of the house (or whichever is the "up the mountain" wall) and take a look at the slope of the land. Does it slope toward or away from the front wall of your house? If you were a rain drop and you landed 5 feet from the front wall of your home, where would you go? If your answer is "it will stream toward the house", then this is the place to start improving the situation. Somehow you need to arrange the dirt so that it either leads the water away, or put in drains that will collect the water and lead it away.

Now look at your gutters. No gutters? This may be ok if you have a four foot roof overhang and excellent slope grading, but here in the mountains this is rare. You'd be amazed to know how much water runs off a roof in a rain storm – if you have a typical 1500 square foot roof, and it rains one inch an hour for 2 hours, you will have almost 2,000 gallons of water running off your roof. This is the equivalent of taking eighty (80) 10 minute showers (with a 2.5 gallons per minute showerhead). If all this runoff lands at the side of your foundation wall on the upslope – it will eventually find its way into your basement.

So, make sure your gutters are clean – even if you have "gutter guards" – small debris can still get past them – and take a look at your downspouts and the drains they run in to – sometimes the connections break and the downspout will deposit water outside the drain. Also make sure the drains are not clogged – you can check this with a hose – run it into the drain; if it backs up, the drains need to be cleaned.

Lastly, take a look at the front wall of your home (or the upslope wall) and the other 3 sides. How high from the ground is your siding? It should be at least 6-8 inches higher than the ground. The reason for this is that you do not want moisture from the ground to get into your siding (which is usually wood) and rot it; you also do not want termites bridging the gap from the soil to your siding or the wood sheathing underneath. Do you have plantings at the sides of your home? Plants hold moisture, so consider either keeping these trimmed back, or moving them back slightly from the walls.

Now that you've solved your mold problem from the outside, we'll move to the inside solutions next week.

Your Healthy Home: Mold in the Basement Part 2

Last week we talked about the causes for mold developing in your basement – from the outside of the home. It is important to correct the problem from the outside first, or you'll be constantly fighting an "uphill" battle which will never end.

Now that you've pulled the wet mulch away from your foundation walls, cleaned the gutters and fixed the downspouts, checked the drains, cut the vines back from growing into your siding, and put a culvert in, now what? Your basement won't be wet from the outside conditions, but could water or moisture be coming from anywhere else? Let's look.

Go into your basement. Start in one area and methodically go into each and every room looking for damp or wet carpet, water stains on ceilings, walls, or floor; stains or wetness around the base of toilets; look under the sink vanity and make sure its dry; check all baseboards for dampness; and look at your air handler if it's in the basement for any condensate backup or any dampness in the overflow pan on the floor.

Of course if you find leaking toilet seals, fixtures, or such, you should repair them. After checking this off your checklist, now take a good look around for mildew and mold. If you do identify it, how much is there? Mold is like an iceberg – once you start seeing any part of it, there is usually a much larger patch of it hidden behind baseboard, tile, or carpet.

As you begin to identify the areas, use your judgment to determine if you've got a "big problem" or a "little problem" – if the areas are small and confined, then addressing the humidity and ventilation in the basement and a little clean up may be all you need to get back on track; if you start pulling off sections of baseboard and see a lot more mold, or your carpet is wet and smelly in a large area, then you've got a larger problem and should call a professional. Some people are sensitive to molds. For these people, exposure to molds can cause symptoms such as nasal stuffiness, eye irritation, wheezing, or skin irritation. Individuals with serious allergies to molds will have more severe reactions. It is also possible to discover the infamous "black toxic mold" which is the most serious in terms of its effect on people. An excellent site with balanced and accurate information on mold can be found at the Centers for Disease Control: http://www.cdc.gov/mold.

Ok, you've found a little mildew that seems confined to one or two areas, and the basement smells ok. What next? You've solved all your source problems and now you need to get rid of it and keep it away. Next week we'll talk about how to clean mildew and mold off various surfaces, so stay tuned. In the meantime, if you have not already measured your basement humidity, you should do that and turn up your de-humidifier so that the RH (Relative Humidity) is no more than 55%. Run at least one fan in the basement to circulate the air. You should try to keep the humidity at 35 – 50% to stop mold spores in their tracks and prevent damage and deterioration to the items you have in the basement.

Your Healthy Home: How to Clean Mold

Have you found mold in your house? What should you do? We've discussed the importance of solving your water intrusion problems first; now you need to get rid of the mold that's there and keep it away. Here's what to do.

Do you have a large amount of mold, or is it confined to a few small areas? Moisture seeping through concrete walls will drive up humidity and cause mold growth on or in walls, carpeting and materials stored in the basement. Mold can grow under cabinets, behind base-boards, inside walls, in carpet padding and under flooring and tile. If what you are seeing is extensive – areas of wet or moldy wall to wall carpet, large black patches on baseboards, or a very moldy smell – then you should call a professional for help. Trying to clean this yourself could be difficult and dangerous. One thing you do not want to do is release mold spores from surfaces so that they float around to other areas in your home. Professional cleaners know how to contain and remove mold, and they wear serious protective gear. Not sure if its mold? Put a few drops of bleach on it – wait 2 minutes - if it lightens up, its mold – if it doesn't, it's plain old fashioned dirt.

If you've found small areas of mold, you can clean it yourself. First, remove any materials from the area that are porous and have mold on them – this includes paper, fabrics, wood, and carpeting. Use your judgment – you may need to throw things out if they cannot be cleaned.

The first thing that people think about when they hear "mold cleanup" is grabbing a jug of bleach. However, bleach is not a universal antidote for mold, particularly when mold has eaten into raw wood surfaces. Bleach is also falling out of favor as a detergent because it's not environmentally friendly (septic systems for example), and folks tend to use too high a concentration, with dangerous health effects.

When starting a cleanup operation using bleach, you should wear protective clothing – rubber gloves, goggles, and a particulate mask. Use no more than a half cup of bleach in a gallon of water. Wet the moldy surfaces and let the solution sit for at least 15 minutes. Then rinse and dry the surfaces off well, and get a fan and de-humidifier going right away. The surfaces we are talking about are tile, painted baseboard, ceramics, metals, and laminates. If you discover mold on bare wood, you can still use your bleach solution, but to avoid spreading the mold you should follow a procedure of wiping the surface and then folding the wipe you are using to a new clean side to prevent spreading the mold spores to the next area. For detailed information on cleaning mold, see: http://www.inspectapedia.com/sickhouse/action.htm or http://www.cdc.gov/mold/cleanup.htm.

When you are through with the cleanup, run your fans and your dehumidifier until the humidity in the area is lower than 50%. You want to make sure that everything dries out thoroughly. Now set the hygrometer you bought in the area and watch it over several days to see how things are going. Eventually you will want to see NO MORE than 55% RH (Relative Humidity) on this instrument at any time in your basement (or anywhere else in your home for that matter). I recommend keeping it at less than 45-50% if you can so you have a safety margin. If the humidity climbs again, guess what – buy another set of gloves, booties, mask and goggles!

More Mold Solutions

The last few weeks have delivered an exceptional amount of rain to our mountain areas. Even homes that are well sealed might begin to exhibit signs of mold growth if the humidity stays high long enough. This is typical in both slab on grade basements as well as homes on crawlspaces. As a home inspector I have seen homes where the humidity was so high in the crawlspace that the carpet on the floor above it was wet. I have even seen water actually traveling through "sealed" crawlspaces.

If you have a case of wet carpet or extensive mold inside your home or crawlspace, you should not try to clean it yourself. You may not have the expertise or tools that the professionals have, and extended exposure to molds trying to clean them can produce serious allergy and health problems.

If you find small areas of mold growth and want to clean it yourself, here's how to do it. You may be surprised at the advice I'm about to give you, as it may contradict what you have read or heard about how to kill mold.

Do not use bleach on mold. There are 3 reasons to not use bleach. The first reason is that fumes from full strength bleach are toxic. Secondly, bleach cannot reach all of the mold growing in porous surfaces, such as wallpaper, wood, and baseboards. Third, the chlorine in bleach can react with mold chemistry to produce fumes. Even if you wear a mask over your nose and mouth, you can end up inhaling a nasty mix that is harmful to your lungs. An exception is minor mold growth in the bathroom on tile and ceramic (non-porous) surfaces. These areas are safe to clean with bleach, but be careful and do not use bleach full strength.

Much safer alternatives include filling a spray bottle with a 50/50 vinegar and water solution and spraying it on the mold. Use rubber gloves and paper towels and throw them out when you are done. Borax also works well – use 1 cup of Borax to a gallon of water. Hydrogen Peroxide also works well on mold as a 3% solution.

Two other things you should check to prevent mold from starting again include the landscape grading around your house and the humidity level inside your house. To check the grading, get your raincoat on in the next hard rain and walk the perimeter of your home. Do you see any rainwater pooling or running against your home? If you do, then once things dry out you will want to re-direct the water path. This may be simple, like adding some soil to specific areas to force the water elsewhere, or more extensive where you need to hire a professional.

To check the humidity in your home, use a meter to measure moisture in the air (called a "hygrometer"), available in your local supermarket or hardware store. If the humidity in the lowest level of your home is over 55% on a continuous basis, you should purchase and install a de-humidifier. You will need to run the condensate (water) line outside or into a drain. Pull the humidity down to 50% or less and you will be able to prevent mold from growing in the first place.

Solving the humidity problem should solve the mold problem. And your home will feel and smell fresher and be a healthier environment to live in.

Four Surprise Places That Produce Mold in Your Home

For individuals who are sensitive to molds, life inside your house can be a frustrating search to identify where allergens are coming from and eliminate them. While it's next to impossible to get rid of every irritant, there are some things that you can do about hidden mold spots in your home. These molds can cause sneezing, a runny nose, and an itchy throat. Here are four places to check and what to do if you find mold or mildew.

- Your washing machine. If you have a front-loading washer, check the rubber seals around the opening. You are likely to find moldy water trapped at the bottom of this seal. What to do: after each load, use a paper towel to wipe the seal dry. If the other family members are not doing this and you find moldy water in this gasket, use a little water and bleach on a paper towel to clean it out. Tip: purchase a box of disposable vinyl gloves for these tasks so that you don't get mold or bleach on your skin.
- The coffee maker. Really? A coffee maker? Indeed, drip type appliances are notorious for building mildew and even mold spots in the bottom of the water tank. This is an area that stays wet with everyday use. When was the last time you cleaned your coffee maker? Don't remember? Ok, me too! What to do:

fill the water reservoir with a 50/50 mix of water and white vinegar. Turn the unit on and allow the mixture to run through the basket into the carafe. When the cycle is half done, turn the unit off and wait 20 minutes. Finish the cycle and then run cool water through it 2 more times to get rid of the vinegar. Do this once every 4-8 weeks and you'll prevent mold from growing. Tip: Clean the basket and carafe immediately after each use, and leave the lid up so that it dries out between cycles.

- The hollow head of your electric toothbrush. The head is designed to channel excess water from the bristles. Over time the wet interior surfaces can generate mold. What to do: Every few weeks, drop the brush head in a small cup of strong mouthwash for 15 minutes. You can also use 3% hydrogen peroxide or vinegar, but you'll need to rinse well to get rid of the peroxide or vinegar aftertaste. Tip: Dry the bristles after every use. Just use a towel or clean tissue. This will reduce water accumulation around the stand too.
- Are you using a dehumidifier in your basement? A portable dehumidifier is a great way to control mold in areas exceeding 60% relative humidity. There's only one problem: the dehumidifier itself can generate mold inside the water tank and on the cooling coils. What to do: Empty the tank regularly. Once a week, use a chlorine or bleach cleaner spray – a couple of pumps – in the emptied tank to reduce mold growth. Tip: When you're not using the unit, make sure you dry out the tank before storing.

Remember, a checklist always helps you remember the little chores that make a difference. House chores are constant – make your lists and spread the tasks out so it doesn't seem as overwhelming.

All Dried Out? Your Guide to Indoor Humidity

While we wait impatiently for spring to arrive, we continue to suffer through static electricity, clinging clothes, and dry skin. Winter temperatures bring with them low indoor humidity and all of the ailments that can accompany it. Dehydration of nasal passages and the upper respiratory tract increases the risk of infection because dry membranes have trouble protecting the airways. Does it have to be this way? No, it doesn't. Here is your guide to indoor humidity and what you can do to improve the comfort level in your home.

Humidity is the amount of water vapor in the air. Relative humidity, which is what hygrometers measure, states the amount of water vapor in the air as a percent of maximum. So, if your humidity meter says "25%," then it means that a lot more water vapor, 75% to be exact, could exist in the air.

Where this gets complicated is in the relationship to temperature and air flow. When we warm the house, the heat removes water vapor, lowering humidity. The colder it is outside, the harder it will be to maintain a comfortable humidity range inside. Cold air has less capacity to retain water vapor. There are two ways to solve the problem.

The first is to purchase a "whole house" humidifier which attaches to your heating system and distributes humid air through your heating ducts. While this solution sounds elegant and simple, it can be expensive and maintenance intensive. If a whole house humidifier is not sized and installed correctly in your home, it can introduce mold into ductwork and leaks into the piping and room where it is installed. If you decide to go this route, you should consult a qualified heating and air conditioning contractor with experience installing these systems. You should have your contractor service the system along with the other HVAC components twice a year.

While whole house humidification is a great solution, you may want to consider other less expensive alternatives first. Start by making sure you know what the humidity is in various areas of the house. You may find that you only have low humidity in one or two areas. If that is the case, you can add an inexpensive room sized humidifier. You might consider a humidifier in the bedroom at night for comfortable sleeping, for example.

Other alternatives to electric humidifiers include clever ideas you may have already tried: shallow pans of water in a sunny window; pans with gravel filled with water under potted plants; boiling water on the stove; letting the steam from your shower out the bathroom door instead of running the vent fan; hanging washed clothes on a line in your laundry area; and leaving the dishwasher door open to air dry dishes instead of using the heated dry cycle.

Consider adding live indoor plants to your home. Plants release moisture vapor, which increases the humidity of the air around them. Plants also filter out chemicals in the air, and release oxygen.

When you take your hygrometer readings in your home, your goal should be 30-40% humidity in the winter-time and 45-60% in the summer months. Less than 25% will be uncomfortable, and more than 60% will encourage mold growth.

One more thought: spring is around the corner!

Do You Need a Humidifier?

Scratchy eyes, dry throat? Does an electric spark jump at you from the TV after you walk across the carpet? Does your winter sweater cling? Aha! No, you're probably not getting the cold that is going around. You are suffering from low humidity.

Too little or too much – humidity is uncomfortable at either end of the scale. According to the American Society of Heating, Refrigerating and Air-Conditioning Engineers (ASHRAE), the ideal humidity range for humans is between 30 and 60 percent relative humidity. Relative humidity ("RH") tells us the percentage of water vapor held in the air. Warmer temperatures can hold more humidity, which explains why we often have too much humidity in the summer and not enough in the winter. The perfect comfort range for people is 45-55% RH.

Last week in the engineering department where I work, I decided to measure just how much water was in the air by placing a hygrometer, or a meter that tells you what the relative humidity is, on my desk. I was amazed to see a reading of 19%. Later at home I did the same thing, and found that it wasn't much better at 24%. What can we do if we want higher humidity and more comfort? Here are some alternatives.

Go natural. Plants release moisture. Place a plant assortment in a sunny area with pans filled with gravel under them. Keep the gravel and the plants watered. This should raise the humidity in the surrounding area by 5-10%. Plants also add oxygen to the air, so you get more than just humidity from plants.

Mechanical ways to introduce more water into the air range from whole house humidifiers that work with your heating system to small units that work in a room. While the whole house units are expensive, the small room sized units are reasonably priced and work well.

Back to my desk at work. I didn't think it would be a good idea to haul in a portable humidifier and plug it in at my desk. Space is tight, electrical outlets are in the next cubicle, and I couldn't see having to take the time to keep filling it with water and cleaning it every week. What to do?

I did some research. After a short time on Amazon reading reviews and looking at specifications I found a nifty little gadget called "The Amazing Humidifier." It's a round cap like device several inches across weighing about 3 ounces with a 4-inch filter that drops down into any container. You can place it in a bottle or a coffee mug filled with water. The cord is on the outside, and plugs into a USB port on your computer. Once powered up, the device soundlessly draws water through the wick-like core, dispensing it into the air as cool vapor at a rate of 30 ml per hour. This isn't much, but it is enough to provide some dry air relief if you've got it going right next to you. You can also plug this thing into a cell phone charger if you're traveling or use it on your bedside table.

Less than a week later, for less than $30, I was sitting at my desk at work breathing in the cool, moist air. Fellow engineers stopped by, wondering what "that thing that is smoking on Lisa's desk" was, and they were "amazed" to find out that it is water mist. I think over half the department has now ordered them.

All About Humidifiers

Even though we've officially entered winter, the unusual warmth has staved off those moments when your sweater makes connections with electrical sockets from five feet away and scratchy nasal passages make us feel like we are catching something. But never fear, low humidity is near.

When we see some seriously cold winter temperatures, some of us may haul our little humidifiers out of the closet. Need another humidifier? Here's your guide to portable units.

First things first. Do you know what the humidity is in your home? If not, you might want to test it. You can buy a hygrometer – a simple device that measures the amount of water vapor in the air. Relative humidity, which is what hygrometers measure, states the amount of water vapor in the air as a percent of maximum. So, if your humidity meter says "30%," then it means that a lot more water vapor, 70% to be exact, could exist in the air.

Where this gets complicated is in the relationship to temperature and air flow. When we warm the house, the heat removes water vapor, lowering humidity. The colder it is outside, the harder it will be to keep moisture in the air inside.

So you take a reading and find out that 2 rooms are at 22% and you feel dried out. What humidifier should you buy?

There are 2 types of humidifiers – Warm Mist and Cool Mist. Warm Mist humidifiers heat water to a boil, then emit the resulting steam. In the winter, they can add warmth as well as moisture to a room. They are not noisy but they do produce a hissing sound. The second type, cool mist, come in three varieties: Evaporative, Ultrasonic, and Impeller. Evaporative models use a fan to blow air over a wet wick. Ultrasonic humidifiers use a vibrating nebulizer to emit water. Impeller models produce mist using a rotating disk. The ultrasonic versions are almost noiseless.

Choose the type depending on the size of the room and how quiet you want it to be. I recommend going to www.consumerreports.org to see the ratings (this magazine does not accept advertising so they are unbiased in their evaluations).

Things to consider when buying a humidifier are:

- Is the unit easy to clean? Bacteria can build up quickly in some models, causing flu-like symptoms and irritation to those with asthma and allergies. All models can emit bacteria into the air, but the cool mist evaporative units emitted the least in lab tests.
- Is the unit easy to fill? Will the tank fit under the faucet?
- Are controls simple? Some units have a humidistat to tell you the humidity and some will shut off after reaching the humidity level you have selected.

Follow the directions carefully on cleaning your humidifier. It's not something we want to do every day, but unless you have a cool mist evaporative type, bacteria will build up. Consult the ratings to see which units are the best in this regard.

In the small room category, the Crane ($45) and Safety First ($30) units did well. In the medium room category, Vicks was at the top at $50; and in the large room category, SPT is at the top at $75. If you want a really big unit, choose a console, where the Essick and Kenmore did well for less than $140.

Aim for between 35% and 50% humidity to achieve the most comfort. Now your sweater won't feel like it is generating power!

14

Dealing with Pests

Summertime: Keep Pests Out of Your Home

Recently a neighbor told me that bats had gotten into his attic. I listened carefully, because I knew that a bat colony gets bigger over time and it's not easy to get rid of them. We think of bats and caves together, but if there isn't a cave handy, your attic will be perfect.

"They had babies." Said my neighbor. Uh oh, I thought, more bad news. If you find out where the bats are coming in and seal the opening, the babies will crawl down inside your walls trying to get out. If you've experienced a dead mouse in your home somewhere, then you know what I am talking about.

"So, what did you do?" I asked.

"I called a professional, and he figured out where they were coming in, and was able to get them, with the babies, out. The key to keeping them out is to put attic vent screening on the outside of the vent, not on the inside."

How do you make sure that bats, mice, squirrels, and bee colonies don't move in with you? Our homes seem tight and secure to us, but a mother bat only needs 3/8ths of an inch of space to enter your home. Here are some tips to keep free loading animals outside.

- Take a slow walk around the outside of your home with a pair of binoculars. Look at ventilation areas under the eaves – the soffits – and make sure that the screens are firmly in place. Now check vents. Some vents – like the dryer vent – should not be covered with screening for an obvious reason – the purpose of this vent is to exhaust hot air from your dryer – along with lint. This vent should have a good flapper seal, meaning it is closed when the dryer is not operating. If it is missing or is stuck open, you should repair it.
- Go into your attic on a sunny day. You should see light coming from the eaves – the soffits – but not anywhere else unless you have other vents, such as a fan or turbine. Any areas where you see light should be investigated.
- If your home has a crawlspace, make sure there is no way that vermin can gain entry. Check the crawlspace entrance door for gaps. Seal cracks and small openings with steel wool and caulking.
- Next, take a look at any piping or wiring conduits into your house. You'll find these at the electrical meter and where the water pipe enters the home. If you home was built recently, contractors might have forgotten to caulk these.
- Always double check repairs after they are done to make sure holes are sealed up. Electrical and plumbing jobs can leave areas around pipes and conduits where small pests like mice and bugs can enter.
- Check windows and doors for gaps. Over time structures settle and sometimes shrink. Check window screens.
- Finally, look at any shrubbery that is close to entry doors. Amazingly, rodents and bugs have a habit of darting out of bushes and in through open doors. Keep bushes trimmed so you have several feet of clearance. Also trim back tree foliage next to your roof – arboreal rats love swinging from a tree to your roof.

If you find yourself with uninvited guests – from bees to bats – call a pro. They have seen it all and will have the right solution.

All About Carpenter Ants

Last week we talked about carpenter bees. "Carpenter what?" I know you said; "You mean Carpenter ants, right?" We did mean carpenter bees, but since you asked, we'll talk about carpenter ants.

Do you see any ants in your house? Do you immediately think you have termites? Of course; we all dread the thought of finding termites, because we know they can cause so much hidden damage to our homes. Carpenter ants and termites can be confused because termites do look like ants to the uneducated eye. A couple of ways for you to tell: termite wings are twice as long as the body, and have a broad waisted appearance (a straight rather than segmented body), and straight antennae. Carpenter ants have segmented bodies, elbowed antennae, and smaller wings. You can't just go by the wings, because there are winged and non-winged versions of each pest. It's also interesting to note that the carpenter ant likes eating termites, so you won't find them co-habitating in the same area. I'm not sure if this is good or bad. It would be great to find a natural enemy of the termite, but the carpenter ants cause enough damage of their own, and so is not a good candidate for this job.

If you are looking at what you think are ants in your home, you should definitely call a qualified pest control company. If you think they are carpenter ants and do some simple baiting or spraying, and the little guys turn out to be termites . . . you get the picture.

So, what about the carpenter ants? Like carpenter bees, they do not eat wood (like termites do). But they do chew through the wood to make their nests, just like the carpenter bees, and can do a lot of damage, including structural damage, to your home. Do you see a trail of sawdust on your deck under an eave? Then you probably have carpenter bees or ants.

Carpenter ants are among the largest ants found in buildings. Winged male and female carpenter ants (known as swarmers) emerge from colonies from early Spring through mid-Summer. Although they like soft, moist wood outside, they will find ways into your home if they think there is food to be consumed. Nests can contain as many as 4,000 ants, and they make a distinct rustling noise inside your home. You weren't "hearing things" after all – it's the ants!

Once the pest control company has determined that your ants are carpenter ants, and applied professional baits, you should control them in the future by doing the following: keep tree branches trimmed back from your roof (the ants use the branches to reach your house); check your house and trim for any moisture or wood rot and get it fixed; keep the gutters clean and free-flowing (so as to not hold moisture against the trim); and do not stack firewood next to your house (will be hard to see your siding to inspect it, and will invite termites), and check seals around doors and windows. Carpenter ants (and termites) love moisture, especially wet wood, so keep everything dry and well maintained. No more invitations to enter your home! Now you know.

What Are Carpenter Bees

I was inspecting a beautiful cedar framed home on Lake Chatuge recently when I heard a buzzing noise and looked up from the side deck to see several large bees coming right for me. They flew straight at my head and at the last minute I ducked as they flew by. As I looked down, I noticed a line of fresh sawdust on the deck directly below the roofline.

"Carpenter bees!" I exclaimed, moving away from the roof fascia board above, where I knew mother bees were boring fresh nests in which to lay their eggs.

"Carpenter what?!" said the home buyer as she came around the corner of the house and up to where I was inspecting 3 perfectly round, exactly one half inch diameter, holes in the fascia trim of her new home.

"Carpenter bees," I said. "It's springtime, and the mother bee bores a hole into your wood trim – only the horizontal members by the way – and lays her eggs. After going into the wood one inch, she turns 90 degrees and moves down the board inside 4 to 6 inches – some tunnels found have been 10 feet long – and then she creates individual nesting cells for her baby bees. "

The 2 furious bees that had dive bombed me earlier were now back, buzzing our heads. We moved away from the roofline.

"The male bees are highly territorial, and stay buzzing around the nests in protection, but the males don't have stingers, so they are scary but harmless," I continued.

"That's very interesting," the homebuyer said, "But look at my trim!"

We both looked up and saw multiple lines of holes, and 2 open tunnels about 6 inches long. This was repeated on the other sides of the home in various degrees, and some of the unfinished wood was chewed or scratched by woodpeckers trying to access the nests for a meal.

"What can be done?" asked the concerned buyer.

"Well," I replied, "Carpenter bees are definitely annoying and damaging, but rarely pose a structural problem unless the damage has gone on unchecked. The bees do not actually eat the wood – unlike termites – which is good, but the bad news is that carpenter bees are hard to control for this very reason – adding pesticide to the wood does not help much. Your best bet is to have a qualified pest control company take a look, and along with the routine inspections for other pests they can spray these nests with boric acid dust or other appropriate chemical – then you should caulk up these holes and tunnels and make repairs. Some pest professionals say you can take care of these yourself by spraying WD-40 into the holes and then caulking them. Apparently, the bees do not like oil based surfaces."

"Well My Goodness!" said the curiously intrigued homebuyer, wrinkling her nose as she observed a female bee entering one of the perfect half inch holes in her roof trim. "What can be done long term to discourage this?"

"About the only thing you can do is stay on top of repairs and keep a heavy coat of varnish or paint on these surfaces," I said. "And if you decide to replace the siding, fascia, or deck rails, try to find a good synthetic product – plastic or composite – that will eliminate the problem entirely!"

Carpenter Bees – now you know!

More on Carpenter Bees

Every year about this time I get at least one question from a Progress reader about carpenter bees. Bees you say, carpenter bees? Don't you really mean carpenter ants? No, I do mean bees.

Those of you who have seen, or are seeing, little piles of sawdust on your deck know what I'm talking about. Once you notice the fresh sawdust and wonder where it is coming from, you notice the bees. Male bees will be flying aggressively, trying to make you leave the areas where the females are boring into the wood. The males do not have stingers, so they can't bite you. However, you would not know that to hear them buzzing around your head. Finally, you look up and notice one quarter to one half inch perfect round holes scattered about on the fascia of your house.

After going into the wood one inch, the female bee turns 90 degrees and moves down inside the board 4 to 6 inches – some tunnels found have been 10 feet long – and then she creates individual nesting cells for her baby bees. This not only causes damage to your fascia boards, it also attracts woodpeckers. The woodpeckers think the baby bees are a wonderful gourmet treat. The result can be house trim that not only looks like it was eaten from the inside out – it was – but also as if someone took a rake and tried to take the trim off at the facing.

The bees do not actually eat the wood – unlike termites – but the bad news is that carpenter bees are hard to control for this very reason. Adding pesticide to the wood does not help much.

What should you do? You have several choices. You can ask your pest control company to take a look, and along with the routine inspections for other pests they can spray these nests with boric acid dust or other appropriate chemical. Or, you can take them on yourself and spray WD-40 in the holes. Caulk them immediately to force the demise of whatever is in the tunnel. Apparently, the bees do not react well to oil based surfaces. Then you can proceed with repairs if there's extensive damage.

Over the long term the carpenter bees will return. To discourage them, keep a heavy coat of varnish or paint on the wood surfaces of your home. Good coats of varnish or paint trick the bees into thinking that the surface is not wood. The other thing to consider is replacing your wood trim with cement based or composite products that the bees do not like. And while cedar tends to be an excellent insect and moisture resistant product, the carpenter bees love it as much as the other wood types. Carpenter bees can do damage and force you to make repairs, but they rarely cause any serious structural damage unless they have been using the home for nesting unchecked for a long time.

The fresh sawdust mystery is solved!

All About Termites

Earlier this summer we talked about bees and ants damaging your home. I received the question, "Are you going to talk more about termites?"

Information about termites would take up this entire newspaper, and then some. So let's hit the highlights – the most important things that you need to know as a homeowner. Here in the mountains we have Eastern Subterranean Termites, which live underground in the soil, and are the most widespread and destructive group of termites in the United States and can include the aggressive Formosan species. Termites cause over two billion dollars of damage a year to our homes. They have been around since the time of the dinosaurs. They are voracious wood eaters (a Formosan colony can consume up to 1,000 pounds of wood a year) and did you know they also eat wallpaper and plastics? They can destroy building foundations, wooden support beams, plastic plumbing pipes, sub-flooring, and insulation. Winged adult termites, called swarmers, look like flying ants and emerge from the colonies on flights to search for more nesting areas during the spring. A single colony can contain 2 million members. No wonder a chill of fear spreads through you when you hear the word "termite." Yes, they are like little monsters. Last week we were concerned about carpenter ants, but termites are the king and queens of destruction.

When I inspect a home, I always advise the buyer to get a termite inspection from a qualified pest control company, as well as a thorough general pest inspection. For the small fee that these companies charge, you will be well served by identifying serious problems now and treating your home to keep pests from damaging it. Not only do you want termites identified and treated, but our spring and summer weather in the mountains brings in mice, spiders, scorpions, and an assortment of other walking and crawling pests. I don't know about you, but I would rather not meet ANY of these creatures inside my home.

The steps to take to keep termites away from your home also work well for other pests. Think dampness and access. First, take a walk around the outside of your home. Where is the soil level? Is it at least 6-8 inches below your wood siding? If it's closer than that, you are inviting termites in for a meal. Is the soil damp near the house? If it is, your gutters are not doing their job and should be repaired. Is there any wood stored alongside the house? You should put this somewhere else. Termites and other creatures love wood and wet conditions. Now take a look at the slope of your soil (called "grade") around the house. It should be sloping away – at least one inch per foot for at least six feet – to keep water from pooling near your foundation wall. On the upslope of your mountain home you should have drains or a swale to keep water away. Dry soil next to your foundation will also mean your basement or crawlspace will stay dry.

Now look around windows and doors, inside and outside. Be careful; if you're not comfortable climbing up to see around windows and look at eave areas, ask a handyperson to do this. You want to make sure that doors and windows are fully sealed and caulked – as well as all areas of your foundation wall – to make sure nothing can get in to your home. You'd be surprised what I find on inspections – sometimes there is a one inch gap at the bottom of the main entry door to the house. We go in and out and never look – meanwhile creatures have an open invitation to the inside of your home.

Remove dampness and access, combine with a periodic professional evaluation, and you will be able to sleep at night without the termite monster nightmare!

What You Should Know About Termites

Termites are little monsters. They have been on the earth since the time of the dinosaurs. They cause billions of dollars of damage every year to our homes. Here's what you need to know about termites if you own a home, and when you are buying or selling a home.

Here in the mountains we have Eastern Subterranean Termites. These little nasties live underground in the soil, and are the most destructive group of termites in the United States. This group includes the very aggressive Formosan termite. They are voracious wood eaters (a Formosan colony can consume up to 1,000 pounds of wood a year) and eat some surprising things, like plastic plumbing pipes, sub-flooring, and insulation.

Here are some rules to observe to keep termites out of your home.

- Rule Number One: Get a termite inspection from a qualified pest control company every single year. They will give you a written warranty that certifies that your property has been inspected, and if termites are identified within the contract period, they will treat your house at no charge.
- Rule Number Two: If you are buying or selling a home, make sure that a termite inspection has been conducted by a licensed pest control company. Your real estate agent will suggest this as part of the due diligence checklist. Make sure it is done. The general condition inspection, conducted by a state licensed home inspector, does NOT include a termite inspection. Home inspectors are not allowed to conduct termite inspections in North Carolina because they are not licensed pest control companies. Confusion over this responsibility has caused financial hardship to countless new homeowners. Although 99% of realtors understand this, you, as the buyer, are ultimately going to be responsible.
- Rule Number Three: Eliminate water leaks and wet areas in and around your home. Termites love dampness. Take a walk around the outside of your home. Where is the soil level? Is it at least 6-8 inches below your wood siding? If it's closer than that, you may be inviting termites in for a meal. Is the soil damp near the house? If it is, your gutters are not doing their job and should be repaired. Is there any wood stored alongside the house? You should put this somewhere else. Termites and other creatures love wood and wet conditions (brown recluse spiders are common in wood stacks). Now take a look at the slope of your soil (called "grade") around the house. It should

be sloping away – at least one inch per foot for at least six feet – to keep water from pooling near your foundation wall. On the upslope of your mountain home you should have drains or a swale to keep water away. Dry soil next to your foundation will also mean your basement or crawlspace will be drier.

- Rule Number Four: Seal up areas where pests can enter your home. This goes for all unwanted pests such as spiders, beetles, scorpions, snakes . . . uh, I hate even talking about it.

Remove dampness and access, combine with a periodic professional evaluation, and you will be able to sleep at night without the termite monster nightmare!

Reader Questions on Termites

The last column on termites generated multiple questions from readers. The questions were good ones and I'm going to answer them here in case you were also asking yourself the same ones.

"Please explain more about termite warranties, and should I get one?"

The answer is yes, you should get one (and no, I don't sell them!). There are two reasons for this answer. The first one is that you don't want the expensive repairs and panic that come with finding a termite colony inside your home. Secondly, nearly all homeowner insurance policies DO NOT cover damages caused by termites. Why? Because insurance covers things that are sudden and accidental. Termite damage is, in their eyes, preventable through homeowner maintenance, professional inspections, and forethought.

"What happens if I am having a home built?"

North Carolina's building code requires that all new residences have some type of termite protection applied during construction.

The method of termite protection must be one that has been approved by the North Carolina Department of Agriculture and Consumer Services – Structural Pest Control Division. Your builder's checklist will include ordering this treatment. Once you move into your home, the company that performed your construction termite treatment may also offer you a termite control warranty. You do not have to use this same company, but I highly recommend that you do engage one of the many licensed professional pest control companies to provide ongoing protection for your home.

Before providing you with a warranty, the company will treat your property to the minimum level required by the state. The cost to initiate the contract varies and you should check around locally to find out what these up-front costs are. Once this has been done, your warranty will be renewable yearly with a fee and re-inspection of your property.

"How long is a termite inspection valid when buying a home?"

A "Wood-Destroying Organisms Report" (WDO) is provided to the buyer and lender at the time of closing. To be valid, the termite inspection must typically take place no longer than 30 days before closing. Depending on the state, the report may be valid for up to 90 days after closing. This agreement means that if termites are found in the structure inside that time period, the provider of the letter is responsible for the repair and treatment – depending on the rules and regulations of the Department of Agriculture.

"Why can't my home inspector also look for termites?"

Although your home inspector has had some training in recognizing visible and obvious pest infestations, if they are not also a licensed PCO, or Pest Control Operator, they are not qualified to inspect or treat a home for termites. Few home inspectors are also PCOs but some home inspection companies also employ licensed pest control operators as part of their services.

Please take my word for it – don't shortcut the termite inspection. This is money well spent. As a home inspector I have seen termites do extreme amounts of damage. A handy site with questions and answers is the North Carolina Department of Agriculture and Consumer Services page at www.ncagr.gov. Type in "structural pest" and you'll find everything you ever wanted to know about termites.

Keep Mice Out of Your House

Mountains, streams, and forests. Our most lovely habitat is also frequented with creatures that want to move in to our homes. They will test our patience trying to get in, and even when we think we've sealed them out, there they are once again. We've all got a story about squirrels, bats, raccoons, and . . . mice.

Oh, the cute little mouse! As cute as they are, they can be dangerous: mice harbor infectious diseases that can be transmitted to humans, including the serious Hantavirus. The best policy is to not let them in.

This is easier said than done. Mice get into our homes easily. They laugh at us as they squeeze through holes the size of a pencil eraser. They scurry about at night, collecting food bits from our cupboards and nesting material from our closets. Think you don't have any mice? Leave some bird seed and peanut butter out on the counter in a saucer with some talcum powder on the periphery and see if there are tracks the next morning.

It doesn't have to be this way. Here are tips for keeping mice out.

Seal them out. Take a walk around your home. Walk slowly and examine the foundation sill line that is just under the lowest line of siding. This is a popular entry point and you may find gaps of an inch or more in some spots. Seal these with silicone caulking or fine mesh screen. Repair windows and doors so that there are no gaps wider than an 1/8 inch.

Inspect around pipes and lines entering the home. Look at vents, electrical entries, gas lines, and HVAC lines. Look everywhere – mice can climb most anything, and jump up to a foot.

It is the time of year when we love seeing the many colorful songbirds that come to our feeders. One local pest control professional said to me, "Do you want mice around your house?" "Of course not," I replied. "Then you better get rid of your bird feeder," he said. "Oh no! I can't get rid of the bird feeder!" I replied, aghast. "Mice love your birdseed, they will hang out." He replied. "That's a tough choice." I said.

If you're like me, and can't stand to not have the birds, then sealing up the outside of your house is the best thing you can do, along with keeping your seed in a place where the mice won't raid it.

What should you do if you have a mouse problem? You can call a professional pest control company, or you can put out traps yourself. You can trap and let loose (not just outside your front door, but considerably away from the house), or kill them with spring traps, glue traps, or poison traps. It is very important to remember that if you have children or animals in the house you will need to be careful that they cannot enter the area with the traps. You do not want children or pets handling traps or mice.

Remember that mice can also sneak in through open doors, in bags of produce, and run in to the garage when the door is open. If you don't have pets or kids, keep some sticky traps out in the garage to catch these occasional freeloaders.

Ladybug Tricks

Hard to believe, but spring really is right around the corner. We know this because the ladybugs are getting through storm windows, regular windows, screens, and other impenetrable objects to land inert on our floors and window sills. I think this may in fact be some alien species that is able to disappear, like the Star Trek crew in a transporter, and then suddenly re-energize inside our room spaces.

What is actually happening is that the ladybugs, or the "multi-colored Asian Lady Beetle," - its formal name - is waking up inside your home's cracks and crevices, having found these places as winter approached last year. With warm weather approaching the beetles will be trying to get back out.

The beetles are actually good for the ecosystem, if not for our picnics. They eat aphids and scale insects that are injurious to trees, shrubs, and agricultural crops. During the spring and summer, these beetles consume large numbers of plant-feeding pests, thereby reducing the need for pesticides.

However wonderful the beetles are, so cute and pretty, they are a nuisance when we are trying to have our picnic, and a nuisance when they find their way into our homes. In the fall especially, these little creatures can be maddening as they fly into your face, nip your arm, swim in your coca cola, and crunch under your feet. If you crush or swat the little darlings, they exude a nasty smelling fluid that will stain whatever the fluid lands on. The creatures may cause allergic reactions in humans as well.

What can you do? Here are some suggestions.

- Keep the bugs from getting in to the house. The same things that you do to reduce energy consumption – sealing cracks – works well to keep pests out. Seal cracks around windows, doors, siding, utility pipes, and vents. Use weather stripping or a good quality caulk to do this. For large gaps use "Great Stuff" sealant, or foam stuffed into the crack followed by a sealant. Inspect doors for cracks and gaps, especially at the threshold.
- Consider placing mesh screening over attic and exhaust vents. If you do this, you should inspect the screening periodically to make sure it's not clogged.
- If you enjoy leaving windows open a lot, inspect the screening carefully and repair or replace. Check for any gaps around the window and frame.

- Avoid using a chemical attack on ladybugs indoors. Foggers and sprays will not reach the ladybugs in their hiding spots, and will add to the amount of contaminants already in the room. If you have a large indoor infestation, call a professional. When it comes to reducing the inside pest population, my favorite tactic is to spray outside around doors and windows and the area where the foundation meets the siding all around the house.
- Vacuum them up and throw out the bag immediately because they will start to smell. Or, if you'd like to release the hibernating ones outdoors (don't laugh, remember these beetles are good for your garden), use a clean vacuum bag and release outside. You can actually keep them in the bag with a moist cloth in a cold place all winter and they will be fine when you let them out in the spring.

With the kind of winter we just had, we long for warm, sunny weather. Guess what – the rest of the creatures in the landscape are waiting too. We'll be ready.

15

Landscaping and Grading

Should You Plan Your Home Landscape?

Every year at this time I go wild planting things. I don't know why this happens, but I suspect that it might be some kind of primordial urge that drives me to dig around in the dirt and mulch. The spring day is lovely, the sun warm on my back, and the soil smells rich. When I get to the garden center, I enjoy all of the color and texture so much that I fill my little VW with flowers and shrubs. When I get home I then convince my husband to return with the pickup truck for trees.

I suppose there is nothing wrong with this planting abandon and you might say, "Have at it." But there is a problem with this fervor; it is called a lack of planning.

As I stood in the driveway yesterday and looked at the house and yard, I realized that there was no rhyme or reason to my efforts. My husband joined me and said, "That Green Giant is too close to the house. Do you realize that it will grow to 25 feet tall?"

"Yeah, right."

"And that little oak tree you planted is too close to the road," he added.

"Uh oh. Well it's still small enough to transplant. And now that you mention it, there's sort of a void over there by the well rock."

We built a little cottage on a half-acre and thought that we'd save some money by putting in our own landscape. Good concept in theory, but I had not thought it all the way out.

Do you have a plan? Whether you've lived in the same place for 30 years, or just moved in to a new house on a bare lot, there are things to consider when you want the landscape to be both practical and pretty. There are three ways to go when you want a cohesive plan to enhance your home's value and make you smile every time you drive in.

- Hire a professional planner. If you have the bucks and not the time, this is the way to go. Expect to pay 20 to 30 percent of the home's value on professional planning and installation of trees, shrubs, water features, and flowers.
- Consult the experts and have them give you several suggested plans that you can install yourself. This will save you time and money, but be ready to invest your own effort on the install.
- Do it all yourself. This is ambitious, but if you have the time and a passion for planting, it is a lot of fun. Don't do what I did. Before you run to the garden center:
    - Research, research, research. What hardiness zone are we in (Clay County is 6)? What kind of soil do you have? Draw out the house and other existing plants and make copies so you can try different designs on paper first.
    - Drive around local neighborhoods and photograph what you like.
    - Spend a lot of time at local nurseries and garden centers and talk to the people there. Good thing they don't charge admission.
    - Get online or to the library and read up on local plants and landscape design. If you love planting, you will love the research.

Don't be afraid to make a mistake. You can always correct something, unless you leave that Green Giant in too long next to the house like I almost did.

A great landscape will add richly your home's resale value and your enjoyment. No matter what method you choose, don't leave landscaping out of your home improvement plan.

Do It Yourself Landscapes

If you've ever paid a builder to construct your home, then you know that the choices are mind boggling. From interior colors to choosing appliances, you have a lot of responsibility for final results. Many builders do not include landscaping when quoting you a price for your home. This is because the landscape is not usually a builder forte, and it's not inexpensive. In the last column I suggested that you not skimp on landscaping because a great design well executed will add thousands of dollars to resale value. More importantly, an interesting landscape will add enjoyment and pleasure to your life in the home.

Can you design and install your own landscape? Absolutely. Is it a big job? Yes, depending on what size your lot is, and if there is anything there now or not. Don't make the decision lightly, though – design is time consuming, and installation more so. Professionals typically charge $15K to $50K for a 2,000-square foot home with an acre lot. But if the idea of doing it yourself gets you charged up, then read on and I'll hit the highlights of how to start and what to do.

Look at property again. Go back to the ones that featured landscapes you liked. Make notes about what you liked – tree lined entry? Inviting patio with gravel paths? Cozy back yard?

Now go online and look at landscapes. The site Houzz.com is a great place to begin. Let the ideas seep in. Spend some time brainstorming a wish list. Engage the whole family, and entertain all ideas.

Some of the items might include a flower lined pebble walkway, shade trees, an expanse of lawn with curved borders, a fence or water feature.

Take a notebook into the yard with you and free hand some sketches. It doesn't have to be pretty. Have someone help you measure the general areas in your drawings so you have some perspective for your design. If you enjoy computer software, there are landscape design programs online, and tons of advice. Look at The-Landscape-Design-Site.com for a start.

Come up with several designs based on your budget. Consider staging the project so that you do the installation in increments to lower the initial cost. Consider breaking the designs in to two separate projects – the front yard and the back yard. They will be quite different, since the front yard revolves around the entry to the home, and the back yard revolves around the personal use of space, relaxing, and entertaining.

Once you have your design, how do you choose specific plants? If you're not intent on becoming a horticultural expert, find the top recommended local nursery and go there. Tell them what you are trying to do. They will be happy to help you choose plants, and they can also obtain plants they might not have.

When should you install your landscape? Now! It's the rainy season and plants are dormant. The weather is not conducive to planting, and it's more of an effort to find the plants, but the professionals will tell you that the fall and winter are good times to start. Another bonus is that if you realize you need help with the project, it will be easier to find in the winter. If you start the design yourself and run out of steam, you can always hand it off to a landscape crew and still have what you want.

Why You Need Landscape Contractors

Confusion abounds when it comes to figuring out what a landscape architect, a landscape designer, a landscape gardener, and a lawn maintenance person does. While the architect and the designer both deal with what goes where, the gardener and maintenance people install and maintain landscapes. Landscaping includes pathways, fences, stonework, fountains and streams in addition to plants. I'll explain the differences so you can get what you expect.

Real estate professionals will tell you that a great landscape design with healthy and well-chosen plants will add up to 15% of the home's value to resale. Yet, home buyers who decide to build a house often leave this component out due to the expense. Even a modest $150K bungalow will require $15K of landscaping. It's easy to skip this part of a new home, but a big mistake that will only grow as the years go by.

If you are building a home, your builder should give you several options on your landscape creation. Don't skip it. Smartly designed and installed landscapes will grow in value over time. If plants and trees are installed haphazardly, the value will decrease instead of increase. Set aside 8-12% of the build price of the home for landscaping. Depending upon the type and price of the home, consider the following professionals.

Landscape Architect. Plan on spending 15% of your landscape budget for a licensed and registered architect if your home is large, expensive, or sprawling. Why? Because the investment that you are already making is substantial, and you will want a cohesive design that substantially adds to the property's value. A great design will integrate what you want as an owner with correct horticultural, spatial, and grading details.

Landscape Designer. A landscape designer does not have to be licensed but may be certified in a variety of knowledge areas. The experience level of a designer is very important since there are few regulations surrounding the profession. Choosing a designer who is well thought of in the community can be an excellent choice for less expensive homes. Designers charge less because they do not have the level of education or licensing that architects do. And while landscape architects do not themselves install the landscape, many designers do the install also. Both architects and designers will present you with detailed color plans, sometimes in 3D, with plant and materials lists.

Landscape Gardener. Gardeners install and maintain landscapes. Gardeners have substantial horticultural knowledge and can determine when something is going wrong and correct it. They take care of everything from irrigation to fertilization to keeping everything the right size and shape.

Landscape Maintenance Professionals. Sometimes called "lawn maintenance" people, these folks keep everything fertilized and neat. Some maintenance firms plant and consult as well, and some do landscape features such as walkways, stonework, and water fountains.

When hiring any contractor, do your research. Ask for references and then find other customers who have used the contractor. Conduct a thorough interview before hiring them. If you're hiring them for a large job, ask for a contract with staged payments.

Finally, don't skimp on the landscape if you are having a custom home built. Fifteen years from now when pathways flow to intimate spaces with color and trees shade the winding drive, you won't be sorry.

Adding a Shed to Your Yard

Last year my husband and I were driving past a store that sold ready to install sheds. At the same time, we were designing a small cabin. I looked carefully at some of the sheds, which appeared to only need an electrical hookup and plumbing to "move in."

"These are pretty fancy looking," said my husband.

"I'll say," I replied, "Look, there's one with a second floor and window curtains."

"Ok, scrap the cabin plans, we'll save some money."

While the building inspector is not likely to approve your new 120 square foot "shed cabin" as a residence, these buildings are very convenient as you discover that your garage or utility room just won't fit any more weed eaters, lawn mowers, blowers, tractors, carts, rakes, shovels, wheelbarrows, children's toys, or patio furniture. Here are some things to consider when deciding what shed to buy.

Permits. In general, most residential building codes do not require permits for "accessory buildings" when they are 12 or less feet in any direction and less than 400 square feet in size, or one story. Interestingly, this does not apply to tree houses that are solely supported by the tree. But since it is unlikely that you are going to store your tractor in a tree, we can leave this large shed-in-a-tree discussion until Christmastime.

Setbacks. You should check with the inspector about how close you can place your shed to your property line. For practical reasons, I'd leave at least 10 feet of clearance. Most zoning requirements specify 5 or 10 feet, depending on where you are placing the structure on your lot.

Tie-downs. North Carolina 2012 residential building code says very little about "accessory buildings," but it does say that your shed should be anchored to the ground. To see detail on anchors and foundations, google R101.2 NC Building Code.

Utilities. If you want electricity or water in your shed, it is no longer an "accessory building," and you will need to get a permit and follow the code rules. As a home inspector I found plenty of "accessory buildings" with extension cords run from the house. Don't do this – you are risking a fire. Things that are illegal usually have good reasoning behind them. If you need utilities, then go ahead and permit them and use a licensed contractor.

Shed design. Once you determine what size shed you need, the rest is up to you. If you want ease of construction consider buying a shed already constructed or a plastic resin shed from the local stores or online. Plastic sheds snap together without tools. If you are handy and want to save money, build the shed yourself. Type "shed plans" in to your favorite browser to see hundreds of free designs.

Shed Taj Mahal. As you begin thinking about shed space, you may decide to build something a little bigger and permit it for utilities. You can go with a prefab or build it. Although it will cost more because of the building requirements, it may be so practical that it saves you money in the long run. A large shed can be a perfect place for a workshop and garden activities.

Small or large, a shed can be a terrific low cost solution to more room with an endless choice of activities and fun.

Waterfall Fun

Are you itching to get outside? Are the sunny weekend days of early spring calling to you? If you need an outside project to relieve all that indoor stress, I have just the thing for you. A waterfall.

If you already live on the water, or have a pool, you may not be excited about this project. If you are not living on the water, then this might tug at your fancy. Waterfall projects can range from very complex and expensive to small and simple. Let's opt for small and simple.

A few homes ago I had a neighbor who really wanted a waterfall in her garden. She told me that it was beyond her ability. I told her that she might be surprised at what she could with just a little help. We sat down at the kitchen table and mapped it out. Several weeks later, a small waterfall graced her garden amidst rocks, lilies, and stepping stones.

Start with location. Because your waterfall needs electricity, you'll need to place the project close to the house or have an electrician run a line to where you want it. We will place our project in a flower or mulch bed next to the house. Make sure the outlet you use is outside rated, and ground fault protected.

Your next decision is one of complexity. Do you want to dig out a small pond and place your waterfall on its edge, or purchase self-contained sections? This all depends on how much you want to spend in effort and money. You can purchase self-contained waterfall units from building supply stores and online. Before you begin you might want to browse the online options and prices and then look around locally. Two interesting online sites are ArtificialRocksFactory.com and Backyardxscapes.com for rocks, and check out Amazon.com for complete fountains. My favorite source is a little-known company called Urdls. In my opinion they have the best selection and the most real looking rocks. To see their online selection, go to www.urdls.net.

If you opt for a self-contained unit, simply run an outdoor type cord from your GFCI protected outlet to the waterfall. Purchase a timer to turn the waterfall off at night. Arrange assorted rocks around your water feature, fill with water, and plug in.

For much less money, you can make the entire waterfall yourself. If you do this, the size and complexity is whatever you decide it should be. Go to the internet and google "creating a waterfall" and follow the directions. You will need to make a small pond around which you will arrange rocks of your choice. Your challenge with this type of waterfall will be keeping the water from escaping. A small waterfall will need about a 120 GPH (gallons per hour) water pump that you place in the bottom of the pond. Run a plastic tube up to the "head" of the waterfall, fill the pond with water, plug in the pump, and make adjustments. You may need to fill the waterfall once a week or more because of natural splashouts.

Use a hollow artificial rock or upside down flower pot to protect your timer and cord plug-in area. You can now enjoy the sound of cascading water in your flower garden which will attract birds and butterflies, take the stress out of a long day, and impress your guests.

16

Home Lighting - Outside

Choose the Right Landscape Lighting

Real estate agents will tell you that a well-developed landscape design around your house can add 10% or more of value to your property. Color, foliage density, and lighting can make your home rich looking. One of the additions that you can make to your yard to help achieve these effects is outdoor lighting.

Path lights illuminate foliage and grounds, creating a magical and inviting nightscape. Tree up-lighting can create drama and interest, and angle lighting into your driveway and landscape can make the yard appear larger. If you have a waterfall garden feature, or are lucky enough to have a creek or pond, landscape lighting in these areas can add an enchanting glow.

To add outdoor accent lighting, first decide whether you will run the lights with house power or with solar power. It may sound like a no-brainer to use solar, but you may want electrically wired light sets for durability and light intensity. You may decide to mix both. Here are the pros and cons for each type.

Wired light sets. The advantages of low voltage electrically powered light sets are that they are bright enough to light large trees, are durable, and they are safe and easy to install. Dozens of designs and types of low voltage lighting are available at your favorite local hardware or home store. Maintenance is low. Safety is high because the wires to the lights plug into a transformer that reduces the electrical feed to 12 volts from 115. Transformers also have a timer so you can set the lights to come when you want, and turn off when you want. This type of lighting does not draw a lot of electricity.

Disadvantages of wired lighting sets are just that – wires! There is nothing complicated – but after you've decided where the lights are going, you need to connect the wiring into snap connectors on each light and run the small cord across the landscape. Where possible, you will want to hide the cord, including trenching it a few inches below any lawn areas. No doubt these lights are more work to put in – but they should then be maintenance free for years.

Solar powered lighting. The obvious advantages here are no wires to connect and no power draw. Durability can be good if you do your research and get high quality units. I recommend going to Amazon.com and reading the reviews by the people who have bought them.

Disadvantages of solar lights include needing regular sunshine exposure to recharge the batteries (8 hours), having to replace the battery periodically, and light intensity that is variable. Solar lights glow rather than illuminate, so they don't work as well for foliage up lighting. Depending on your landscape, these drawbacks may not be an issue, and the convenience might outweigh the drawbacks.

For larger landscapes, mixing the two types of lighting can yield stunning results. Use 12-volt electric lighting to illuminate tree trunks, foliage, and water features. Use solar lighting far from the electric outlets and where it's impractical to run cable.

The first night you see the landscape lights pop on at your house, you will smile and the compliments will flow.

How to Add Lighting to Your Yard

Ready for a fun spring project that will add ambiance and drama to your yard or patio? Landscape lights are simple to install, and will add curb appeal, safety, and security to your home's outdoor spaces. This project can be as simple or complex as you wish, and will add to the resale value of your home. Here's how to begin.

Where would you like lights? You can begin by lighting tree foliage in front of your home, lighting a dark pathway, or making the patio area dazzling. Walk outside your home after dark and take a look around. One light could shine up into the oak tree in the front yard, creating a magical, inviting scene without being bright or glaring. A set of lights can illuminate the front walkway. In the back yard, a few hanging lights in a tree can highlight the fire pit.

Next, draw out where you want light. Start with one area. How many lights? Start small, you can always add more later. On your drawing note where the nearest 120 volt outlet is. You'll need this for the transformer power plug. The transformer will convert line voltage to a safe 12 volt system that is simple and easy to connect. Your only constraint at this point is how long the cord is from the farthest light to the outlet. I have run cord as far as 200 feet, though you may want to stick with what the kit gives you, as it is sized correctly (gauge) for the length.

Next, head to the nearest home store or to Amazon.com. You might want to review all of the landscape kits online before looking in your local store so that you know what you want. Malibu, Paradise, and Moonrays are the common brands. The online reviews of each kit can be helpful.

When you get home with the box, lay out the parts and read the directions. It's straightforward; you'll place the lights where you want them, and then run low voltage cable to each light, ending up at the transformer. Then you will connect the cable, and plug in the transformer. The transformer has an automatic timer and/or a photoelectric switch that will turn the lights off during the day, or according to a schedule that you choose.

Once you have verified that the lights are where you want them and are working, you'll take a spade shovel, make a thin, shallow trench in the grass or mulch, and bury the cord. You can even do this while the power is on.

You may wonder why I don't recommend running high voltage lights, battery powered lights, or solar lights for landscape projects. For the average homeowner, using line voltage (120V) is a safety hazard, batteries go dead fast and are a pain to replace, and consumer grade solar lights usually fail after about 18 months.

The great thing about low voltage landscape lights is that they can add some excitement and fun to the landscape but don't contribute to light pollution when people aren't enjoying them. Fixtures are replaceable and interchangeable as long as you don't exceed the wattage rating on the transformer. And you can add another kit anytime.

There's a landscape light kit out there for every budget. Have fun!

Safety Tips for the Holidays: Lighting

One of my favorite things during the holidays is seeing the multicolored light displays. Being the techy person that I am, I actually enjoy putting the lights up and flipping the switch.

We all plan our holiday lighting, right? We sit down and draw out where the lights are going and what we will need for extension cords, junctions, etc. We read and follow the directions for how many strings of lights can be hooked up end to end. We go to the basement or attic and pull the neatly stored and labeled lighting out of the box . . . whoops . . . did I just hear you say this is not you? You say all of the lights are jumbled together in an unlabeled box and half of the bulbs are burned out? Someone took the extension cords for another family event? Ok, don't worry. Here are some tricks to make this process easier next year.

Outside: instead of searching out the electrical cords you use in your workshop or garage, why not have dedicated sets that you can put away with the lights? This way you won't waste a lot of time searching for them. Make sure these cords are heavy enough for the length of the run and are capable of handling the load of the lights. Most light strings will tell you how many you can place in series. An overloaded cord will trip the circuit it's on in your electrical panel. How do you think I found this out?

Make sure your outdoor outlets are GFCI ("ground fault circuit interrupter") protected so that the outlet will trip before cords overheat or get wet. If you have any doubt, you can buy a standalone GFCI unit to go between your cords and the house plug. You can also buy an inexpensive GFCI outlet tester for these outlets to verify they are working.

When you take the lights down, check for burned out bulbs and other damage. Fix them NOW before you put them away. Label the boxes so that the contents do not get raided during the year. When next year arrives, you will be so pleased with yourself!

Inside: Make sure you protect the cords running all over the floor like octopi from vacuum cleaners, people, and pets. Small area rugs or duct tape will be a big help, but be careful that the cords do not run under furniture feet or other objects that can apply pressure to the cord and damage it. Use the same common sense you used outside: don't place too many lights and cords on any one circuit – oops, there goes the breaker again.

Last but not least – do turn off the grand displays both inside and outside when you leave the house and at night when you go to sleep. This will help protect you from possible fire and save on electricity. Have a bright and colorful holiday.

17

Home Lighting - Inside

Your Lighting Questions Answered

Progress readers are asking questions about the phase out of the old incandescent light bulbs, and about choosing the new types. Here's a guide for you to use.

Readers write: "I'm confused about watts and lumens. It used to be so simple. I know how much light I get with a 100 watt light bulb. What is this 'lumens' thing?"

Think of it this way: watts means energy consumption, and lumens means brightness, or how much light you get from a bulb. We are confused because "watts" is "watt" we grew up with, but lumens is really a much better way to represent how much light we need and expect. Use the information below to help you choose.

- A 100 watt bulb represents about 1,600 lumens. If you buy a CFL (Compact Fluorescent Light bulb, or the ones that look curly) putting out 1,600 lumens, it will only use 26 watts of your electricity dollar. Good deal!
- A 75 watt bulb can produce 1,100 lumens. If you buy a CFL rated at 1,100 lumens, you will be using only 23 watts.

- A 60 watt bulb produces 800 lumens. A CFL producing 800 lumens will use 15 watts.
- A 40 watt bulb produces 450 lumens. A CFL at this rating will use 11 watts.

"What is this 'color temperature' thing?"

You may have bought one of the new bulbs and decided you didn't like it once you put it in your lamp because it wasn't "warm and cozy." You probably bought a "daylight" bulb without realizing it. You can take it back to the store and ask for a "soft white" bulb. Similar to the tube type fluorescent bulbs, the CFLs are "temperature rated" which is measured in "Kelvin." All you really need to know about this system is that the lower the color, or temperature number, the warmer (or softer, yellower) the light. Use the information below to choose the right one.

- 2700K or less – warm white – use in your home and office
- 3000K – use this if you like a little bit whiter, cleaner glow
- 3500K – use this light temperature in your business or retail environment
- 4100K – use this if you 'd like to duplicate an outdoor winter day with an overcast
- 5000K – use this temperature if you'd like your space to look sterile and white

"Are all of the incandescent type bulbs going off the store shelves?"

No. A wide variety of the old bulbs will still be sold. Interestingly, the largest energy wasters – the 150 and 200 watt incandescent bulbs – will still be available. I have not figured this part out. If the government really wants to move us to more efficient products, I would have expected these to be included in the phase-out. You can still save considerable money on your electric bill by replacing these large wattage incandescents with equivalent brightness (lumens) CFL bulbs.

A variety of small bulbs will also continue to be sold. Examples are the 40 watt bulb in your refrigerator, the grow light for your greenhouse plants, night lights, outdoor post lights less than 100 watts, and yellow bug lights. You should still consider putting the higher efficiency bulbs in some of these applications however – particularly the outdoor post lights. CFL bulbs will work well and save you money.

In the next column I'll answer the rest of your questions on lighting.

More Lighting Questions Answered

Progress readers are asking lots of questions about the new light bulb choices. Last week I talked about watts, lumens, and temperature ratings. This week I'll answer the rest of the questions you asked.

"What about the little clear tear drop bulbs in my fan assemblies and chandeliers? Will I still be able to buy them?"

Small specialty bulbs are exempt from the phase out restrictions, so you will continue to find these in stores. In the future, you will be able to replace these with LED (Light Emitting Diode) bulbs and you won't have to drag the stepladder out so often.

"My CFLs (Compact Fluorescent Light – the spiral bulbs) don't seem to last as long as the package says they should. Is this typical or are some brands just poor?"

The older incandescent light bulbs lasted an average of 1,200 hours. Manufacturers of the CFL bulbs claim a life of 10,000 hours – or almost 10 times as long. But in most consumers' experience, this is exaggerated. A more realistic measure of life is about 6,000 hours for a CFL. Installation also affects the life expectancy. CFL bulbs are designed to be used base down, as in a table lamp, with plenty of ventilation. Heat buildup can reduce their lives, even though they produce much less heat than the old bulbs. I routinely load 75 watt equivalent (1,100 lumens using 23 watts) CFLs into ceiling fixtures, and recessed can fixtures to save money. In this application, I doubt I would get 10,000 hours (or "7 years" as they advertise), but I know I'll still obtain a substantial electric bill savings.

"I read that CFLs have mercury in them. Isn't mercury dangerous?"

Only if you break one. Intact, the bulbs are perfectly safe. And they are pretty durable. If you drop and break one however, a very small amount of mercury vapor will escape. Here is what the EPA says to do if you break one of these bulbs:

- Open a window and ventilate the room for 15 minutes; don't stay in the room.
- After ventilation, use cardboard to round up the remains of the bulb.
- Wearing disposable gloves, use a wet paper towel to wipe up the area.
- Put the pieces in a plastic bag and take to Lowe's, Home Depot, or a recycling center.
- Don't vacuum up the remains, or you could spread mercury dust around.

The general opinion in the medical community is that the amount of mercury is so small in these bulbs that breaking one of them does not pose a health risk as long as you follow the instructions above. It may sound extreme, but these guidelines are designed to protect both you, and the environment, by recycling the bulbs (i.e., don't throw in the trash).

"I understand the CFL bulbs and I've bought them, but what is LED?"

LED bulbs are safe, dimmable, durable, light up quickly, exhibit pleasant color temperature, and will last for up to 50 times longer than incandescents (that's more than 15 years). LED technology is the best of all possible worlds . . . BUT (big but) they are still expensive. The good news is that the prices will continue to come down as manufacturing technology improves.

Time to Switch Bulbs

Some time ago I wrote about the demise of the dear old incandescent light bulb. Incandescent bulbs were cheap and offered a warm, inviting light. There wasn't any confusion about what to get. There was only one shelf in the hardware store for interior lighting and everything made sense.

Then in 2007 the government began a seven-year effort to improve bulb efficiency by banning the largest bulbs – 100 and 200 watt – and followed last month by banning the manufacture of 40 and 60 watt old style incandescent bulbs. The edict did not prevent the manufacture of incandescent bulbs, it just said that the bulbs had to be 30% more efficient, which was not possible with the old technology.

People are still in love with the old bulbs. For years the public has been stockpiling incandescent bulbs. I've talked to numerous Progress readers who tell me their attic is full of them.

Is this unfortunate government interference or is it a good thing? While I do not want to get into an argument about what the government should or should not be doing – there are strong opinions on this issue – I will say that the results have been positive with light bulbs. Although the range of choices in bulbs now is overwhelming, the good part of this trend is that the new bulbs will save you lots and lots of money.

Yes, it's time to switch. Here's why.

- The two best energy savings bulbs – CFL ("Compact Fluorescent Light bulbs") and LED ("Light Emitting Diode") have matured and dropped in price. You can get the new bulbs in warm colors, and you can buy the bulbs in dimmable versions.

- CFL and LED are 400% more efficient that the incandescent. This translates into a savings on your electric bill every month. For example, running the old incandescent 60 watt light bulb for 6 hours a day will cost you about $16 a year in electricity. A CFL bulb in the same brightness range will cost you $3.48, and an LED bulb will cost you $2.60 a year.
- CFL and LED run cooler, and the lower wattages for the same amount of light mean there is less chance of something catching on fire.
- CFL and LED last longer. CFL bulbs last 10 times longer than incandescent, and LED bulbs last thirty times longer than an incandescent. What this means is your investment in new technology will pay off over time.

Go into your attic and pull out all those bulbs and take them to the recycling center. Replace burnt out incandescents with CFL bulbs or the newer halogen incandescents. LED bulbs continue to come down in price and soon will be an even better choice overall. If you were to replace all of the light bulbs in the typical home with LED bulbs, the cost savings in electricity would be over $250. Now that is a bright thought!

Old to New Again: Vintage Light Bulbs

Online retailer 1000bulbs.com reports that the sale of vintage light bulbs has exploded in the last five years. Vintage light bulbs? Didn't we just go through withdrawal from incandescent light bulbs? What's going on? What are vintage light bulbs?

Four years ago, I wrote about the "incandescent hoarders" and I even met some of you in Ingles. You said you loved the old light bulbs and you had gone out and purchased boxes of these bulbs and put them in the attic. Well now you can pat yourself on the back, because the old bulbs are "back" – not on the shelves, but in demand. But they are not incandescent!

Technically, the incandescent bulbs that you put in your attic are not the ones most in demand, although what you have does represent the soft, gentle light that the vintage bulbs provide. Tired of the glare from the compact fluorescents and starkness of the early LED bulbs, manufacturers have actually designed bulbs that LOOK like the old bulbs but use newer LED technology so that the bulbs do not use three times the energy like the old bulbs did.

The vintage bulbs are clear or amber, and show the golden glow of filaments in an assortment of shapes and sizes such as stars, spirals, loops, zigzags, and hearts. Sometimes called "Edison bulbs," the vintage styles add warmth and charm to interiors. To see what these look like, and what they cost, visit www.1000bulbs.com and click on "LED Antique Edison Bulbs."

There's good news and bad news. The good news is that these new bulbs are many times more energy efficient than the incandescent bulbs you have stored in your attic. They are also more antique looking, and come in a style for every taste. The bad news is that they are expensive. One bulb in the catalog – "LED –Filament Type – 3.5 Watt – 40 Watt Equal – 2200 Warm Glow – Dimmable – Amber Tinted" is priced at $14.49 for one bulb. But with a 15,000 hour life, that's 15 times longer than your attic bulbs.

This "bad" news is not really bad news. Here's why. If you were running 30 incandescent bulbs of 60 watts in your home you would spend $328 to supply them with electricity over a year. They would last about 1,200 hours each. Remember what a pain it was to change these things when they were in an out of the way fixture?

If you replace all 30 bulbs with 60 watt equivalent LED bulbs, you will now spend $32 a year to run them, and they will last you 50,000 hours. There are 8,765 hours in a year. So, the LED bulbs will last you almost six years compared to just months for incandescents.

My point here is that it makes more sense to spend the money and gradually replace all of your incandescents – even CFLs (half as efficient as LED) with LED lighting. And now the great news is that you can have virtually any kind of bulb you want in an LED version. So, keep your boxes of incandescent bulbs in the attic.

Are You a Light Bulb Hoarder?

A year ago I talked about the changes in light bulb technology and mentioned that the government passed a law in 2007 requiring most screw in light bulbs to use 27% less energy by 2014. Here in the United States, 100 watt incandescent bulbs were phased out of stores last year. The ban on 75 watt bulbs went into effect this past January, and the 60 and 40 watt incandescent bulbs will no longer be available after January, 2014.

Although a noble effort in the name of conservation, many citizens considered the law intrusive and were upset when they found out that their wonderful warm yellow light casting bulbs would no longer be available. This caused many people to go out and buy boxes of the incandescent bulbs in all wattages. It is human nature to want something that you cannot have, and especially when the government says you cannot have it!

If you are one of those incandescent light bulb hoarders, I have a new perspective for you. I hope when you finish reading this, you will feel much better about not being able to buy the old bulbs, and perhaps even decide not to use the 50 boxes of 100 watt bulbs you have stored in the attic.

First of all, this legislation is not confined to the United States. Countries around the world are moving in the same direction, and it's all about energy savings. Think about it: why use old technology that burns through our electrical power dollars? Incandescent bulbs burn very hot and burn out fast. Only 10% of the energy that goes in to an incandescent bulb produces visible light.

Second, the characteristics that made us hate what I call the curly Q light bulbs – known as "CFLs" - slow to turn on, high cost to purchase, not dimmable, and the sterile white light that they cast, have improved. You can buy them in a warm light tone (called light "temperature"), they brighten faster, and the prices have come down considerably.

But the best news is a bulb now being sold that will match the ambiance of the old incandescents but meet the new energy standard. These bulbs are called "halogen incandescent" and look like the old bulbs you are keeping in your attic. They cast a soft warm light, last one and a half times longer than the old ones, and they will save you 25% or more on your lighting electricity usage. Best of all, they do not cost an arm and a leg.

So, gather up all your boxes of the old incandescent bulbs and take them to the teenager down the block to sell on eBay. There are still plenty of people around the country who don't realize there is new technology that will save them money on their lighting bill.

In the next column I'll answer the questions readers have about the confusion over lighting choices and how to spend the least amount of money to get the highest lighting return.

Latest Tech in Christmas Lights

As the leaves finish falling through the cold rains of fall and the morning field frost sparkles in the early freezes, we have the delight of Christmas lights to look forward to in the weeks ahead. The annual ritual for us begins Thanksgiving weekend and we look forward to it like children. Rather than being viewed as a chore, we enjoy the process.

But it is certainly is a process! We wonder if it could be made simpler. Each year the boxes are dragged out of closets and attics and there is always something missing. What happened to that special green extension cord? Where is the light timer? How did all these bulbs stop working? We now have to make a special trip to the hardware store for things we know we put somewhere but can't find.

Ready to move into the world of new lighting technology? Replace those wattage thirsty incandescent bulb strings with power sipping and durable LEDs? Here are some ideas for your consideration.

Go battery powered. Remember the movie "National Lampoon's Christmas Vacation"? Some families, including ours, enjoy this comedy so much we watch it every season. My favorite scene is the one where Chevy Chase has plugged all of the Christmas light extension cords in to plug expanders all plugged into a single outlet controlled by a single switch in the pantry. As family members travel in and out of the room, the light displays outside go on and off, perplexing everyone and ruining the big "aha" moment when they are all supposed to come on. Assuming that your outdoor light displays are somewhat smaller than Chevy's, it is possible to not even use electrical cords. While just a few years ago battery powered light sets were underpowered and battery eating, the new LED sets are efficient enough to run well on batteries if you are just putting up a few strings or outfitting the tree.

Go solar outside. Solar lights, of course, do not need costly batteries. A small panel the size of a hockey puck powers a string of 100 lights. For comparison purposes, a string of 100 LED lights running for 300 hours will cost you about 25 cents in electricity while using the older incandescents will cost you $2.50 in electricity (about ten times more). So while solar powered lights are free, there is a downside – after a few seasons the panels tend to wear out and the sets are expensive to purchase.

Go smartphone. There is a new light on the block, called RGB. Each bulb contains 3 LEDs – red, green, and blue. RGB bulbs will produce up to 16 million light combinations. An electronic controller runs the show and you can run the show from your smartphone with an app. Manufacturers Lumenplay and iTwinkleLight (put in your browser to see their products) report brisk sales. The downside? Of course – cost – the Lumenplay starter set is a whopping $80.

Go Laser Lights. Prices have come down on laser fixtures that project tiny intense pinpoints of light on to your home or yard. They create an image that floats and looks like something you spent hundreds of hours designing. The company is called BlissLights.

Someone will have to talk to Chevy Chase about updating his Christmas Vacation light displays.

18

Painting Tips

Home Painting Secrets

Spring is springing in the mountains and it's time to get outdoors. Does your home need paint? You can call a professional, or you can do it yourself. The reason professional painting companies exist is because they know how to do the job right to get beautiful long lasting results. A good painter is well worth the price he or she charges. You can get excellent results yourself if you are not a professional painter, but it will take you longer and you absolutely must follow every step. In a previous time in my own life, I found this out the hard way. If you do it yourself, follow these tips and tricks.

- Preparation. Have you heard this before? Did you decide to skip steps thinking it would come out ok? Skipping preparation steps will make results much worse, and take longer when you have to do it over. I can't emphasize enough the importance of leaving yourself enough time so that you don't rush the job. Make a checklist and then follow the directions carefully on your paint can. When you go to your local building store for your supplies, seek out a professional and ask them to help you with your checklist.
- Choose the best value. Do your homework before going to the store. Hit the internet and look up the buyer's guides to paint.

My favorite is Consumer Reports. You will find that the best paints are not the most expensive. Save yourself some money. Believe it or not, our local home improvement stores carry the top rated brands. Don't go into the stores without knowing the ratings and pricing or you may leave with a product that lasts half as long as the best ones.

- Clean edges. One of the secrets to a great paint job, inside or out, is getting a clean edge. We have all had the experience of laying down masking tape thinking that everything was fine until we pulled the tape off to discover a wavy mess of wet paint underneath. Use painter's tape – Frogtape and 3M Edgelock are two of the best. Use finger pressure on the tape where paint will meet it to make sure there is a good bond. Don't leave the tape on the wall for days at a time; you may have trouble removing it.
- Coverage. Don't thin your paint unless the directions call for it. Most paints are meant to be applied directly out of the can. When you apply the paint, get plenty of coverage – paint an "M" shape on the surface and then use cross strokes to fill it in. Don't try to save money by under applying. Good coverage will last longer and make cleanup easier.
- Tools for tricky areas. Don't try to use the paint roller for everything. Buy edging tools and angled brushes when you're in the paint store and use them first to cut in and make the rolling a breeze.
- Don't stretch the day. When you get tired you will get sloppy. Stop and place saran wrap over your roller pan and roller, pressing it down lightly on the wet surfaces. Use 2 layers. Tomorrow it will be as fresh as it is today and you can get started again.
- Have fun. Put on your old clothes, put music on, and enjoy your handiwork.

How to Paint Your Garage Floor

In the last column I talked about organizing your garage. Now that everything is in cabinets or hung on the wall, there is plenty of floor space. I know what you're thinking: "She hasn't seen my garage." Well then, there is *some* floor space. If the floor is almost any material except carpet or dirt, you can paint it.

Painting a concrete floor is not a complicated job, but it is arduous and time consuming. Don't let any of the do-it-yourself articles tell you otherwise. In fact, it's a bit like staining your deck or painting your house: preparation is the key to a great result. We've all tried to shortcut painting chores. Ok, maybe you didn't but I did. The results didn't last long.

Although painting your garage floor is not for the faint of heart, if you follow the instructions and use a really good paint, you will be delighted with the results. An epoxy painted floor will repel dirt and oil, be very easy to keep clean, keep the inside of your home cleaner, and your friends, guests, and family will be impressed. You can even add a clear coat that will make your garage look like Jay Leno's garage, minus the Ferrari and Rolls Royce.

We said preparation is number one. You will need to spend considerable time removing everything from your garage that touches the floor. Then mask baseboards, and vacuum or sweep the floor. Use a one part bleach to 3 parts water solution to scrub the dirt off. Use a bristle broom and elbow grease for this step, and be sure to ventilate the area. Wear a mask, goggles, hat, and a $4.99 Tyvek set of coveralls. When the floor has dried completely, use a crack repair kit if you have cracks in your floor.

Now sprinkle some plain water on the floor. Does it absorb in to the concrete quickly or sit on the surface? If it sits on the surface, you will need to prepare the floor for paint by acid etching it. This is a nasty job, since acid etch contains muriatic acid. Put your Tyveks back on along with mask and goggles. Follow the directions exactly and heed the safety advice.

The next most important component is the paint you use. I have used everything from straight out of the can concrete stain to premium epoxy systems. My strong advice is to go with a 100% solids epoxy system. It costs more, but is thicker and lasts much longer. If you're going to go to all the trouble to prepare the floor, why not apply something that will last a long time? It's not any more difficult to mix and apply, and it looks stunning. Add multi-color paint flakes, and you're a pro. My personal favorite is epoxy-coat.com. All of the high-end epoxy systems provide online videos for the steps I gave you above. They provide handy tips, such as buying a pair of spike shoe attachments so you can walk around all day in wet paint and the surface will close in after you – great for tossing the paint flakes.

If this process does not sound appealing you have another choice. Hire a handyperson to do the job and since you now know how it should be done, you can supervise!

19

What You Should Know About Radon

Confused about radon? Get the Facts

Last year on "Who Wants to Be a Millionaire," one of the questions was, "Balloons are sometimes filled with: A. Xenon; B. Proton; C. Radon; or D. CO2." The contestant called her friend and her friend told her the answer was RADON. RADON was not the right answer and the contestant was eliminated from the competition! The answer to this question can only be D. carbon dioxide (CO2), since the other answers are not correct and CO2 IS sometimes used to fill balloons, and even car tires. And yes, you are correct when you say that balloons are also filled with Helium, but that was not one of the answers!

So, what does balloon filling have to do with your home? Well, the fact that a great majority of people do not know what radon gas is could mean that those of us here in the mountains of North Carolina and northern Georgia could be at a higher lung cancer risk from radon. Radon causes more deaths every year (lung cancer) than drunk driving. [from EPA's 2009 A Citizen's Guide to Radon] If you look at the national map on radon concentrations, you will discover that our counties are in the yellow and red zones, meaning we are more likely to find elevated levels of radon in our homes here.

What is radon? Radon comes from the natural radioactive breakdown of uranium in soil, rock, and water, and ends up in the air we breathe. Radon is colorless, odorless, and tasteless. Although radon is found in all types of buildings, we get most of our exposure when we're at home. Radon can exist in the air, and in our water source. Although radon levels vary throughout the United States, radon has been found in every state. Levels can be medium to high in this area of the mountains. Radon tends to accumulate in the lowest parts of our home since it comes from the soil beneath out homes. Radon gas can enter through cracks and crevices, spaces around pipes, or any other unsealed area.

What levels of radon are ok? Radon in the air is measured in "picocuries per liter of air", or "pCi/L". Generally speaking, levels less than 4 pCi/L are considered safe, although if you can reduce the levels to 2 pCi/L or less it's a good idea. The really good news in all of this is that you can TEST for radon, and you can put systems in the home to lower radon levels. Lowering the level of radon may be very simple depending upon the circumstances. If you have a reading of just over 4pCi/L, you may be able to do some simple sealing of cracks around the crawlspace or basement of your home, or add some ventilation, to lower the levels in the home itself.

You can test your home for radon yourself by going to a home improvement store and buying a test kit. Or you can call a home inspector or a professional radon tester, both of whom have been specially certified to test for radon and interpret the results. Test kits in the store are under $50, and a professional will charge $140 or more (they use 2 kits at a time for accuracy), but that also includes advice on what to do if the levels are high.

The last piece of good news is that if you do need to reduce the levels of radon in your home, it is in line with other home maintenance costs. A vent and fan system is usually the first line of defense (after the sealing of cracks). NOTE: If you do your own test and the reading comes out over 4pCi/L, I recommend that you call a professional for another test of your home to confirm your own reading and get advice on what to do next.

You can learn more about radon by going to the Environmental Protection Agency's web site (www.epa.gov) and by going to www.radon.com. Get a copy of the free A Citizen's Guide to Radon from the EPA site or from: www.TestYourRadon.com.

Take the Radon Quiz

How much do you know about radon? Take this quick quiz:

1. True or False: Clay County is in an orange radon zone meaning the average indoor radon level is between 2-4 pCi/L.
2. True or False: Cherokee County is in a red radon zone meaning the average indoor radon level is over 4 pCi/L.
3. True or False: The EPA (Environmental Protection Agency) advises homeowners with radon levels of 4 pCi/L or more to install a ventilation system to exit the radon from the house.
4. True or False: Testing your home for radon is simple and inexpensive.
5. True or False: High radon levels in any home can be reduced to safe levels.

If you answered all "True" then you get an "A" in radon knowledge! If you are scratching your head and don't know what we are talking about, then read on. High radon gas inside homes is the second leading cause of lung cancer behind cigarette smoking.

What is radon? Radon comes from the natural radioactive breakdown of uranium in soil, rock, and water, and ends up in the air we breathe. Radon is colorless, odorless, and tasteless. Although radon is found in all types of buildings, we get most of our exposure when we're at home. Although radon levels vary throughout the United States, radon has been found in every state. Radon tends to accumulate in the lowest parts of our home since it comes from the soil beneath out homes. Radon gas can enter through cracks and crevices, spaces around pipes, or any other unsealed area. Radon levels are higher in the wintertime because of pressure differentials in our home.

What levels of radon are ok? Radon in the air is measured in "picocuries per liter of air", or "pCi/L". Levels less than 4 pCi/L are considered safe, although the EPA says if you can reduce the levels to 2 pCi/L or less it's a good idea. You can test for radon yourself, and you can put systems in the home to lower radon levels (professional "radon mitigators" do this work). Lowering the level of radon may be very simple depending upon the circumstances. You may be able to do some simple sealing of cracks around the crawlspace or basement of your home, or add some ventilation to lower the levels.

You can test your home for radon yourself by going to a home improvement store and buying a test kit. Or you can call a home inspector or a professional radon tester, both of whom have been specially trained to test for radon and interpret the results. Test kits in the store are around $20, and a professional will charge $125, which includes advice on what to do if the levels are high. If you get one from the store, or online, make sure you follow the directions carefully for accurate results.

You can learn more about radon by going to the Environmental Protection Agency's web site (www.epa.gov) and by going to www.radon.com.

Take the Radon Quiz 2

How much do you know about radon? If you are new to the area, you may not know anything about radon gas. And you may have lived here your whole life and not know about radon. I write about it every year, because exposure to radon over many years can cause lung cancer. Take the radon quiz to see how much you know!

- Radon causes twice as many deaths every year than drunk driving in the U.S. True or False? True. Radon causes about 22,000 deaths a year from lung cancer, while drunk driving causes about 11,000 fatalities.
- Testing for radon is complicated and time consuming, and you need a professional to perform the test. True or False? False. You can buy a test kit at your local home improvement store for about $15. The test takes a few days – you simply hang the kit in your home and then send it to a lab.
- The top floor of your home will test the highest for radon because the gas rises. True or False? False. Radon seeps in to the lowest level of your home from the ground, and barriers such as plastic, insulation, and flooring slow it down. Sometimes simply sealing foundation cracks and slab cracks and other leaks into your home will lower it to safe levels.
- Radon gas comes from outer space during the daytime and is similar to radiation from the sun. True or False? False. Radon comes from the natural radioactive breakdown of uranium in soil, rock, and water, and ends up in the air we breathe. Radon is colorless, odorless, and tasteless.
- The amount of radon in the air outside your home is zero. True or False? False. The average outside level is actually .7 pCi/L, or picocuries per liter.
- Radon is found in other areas of the United States, not just the mountains. True or False? True. Although radon levels vary throughout the United States, radon has been found in every state. Levels can be medium to high in this area of the mountains.

- Clay County radon risk levels are low to medium. True or False? False. Whereas the national average for indoor radon levels is 1.3 pCi/L, Clay County is in a "red" zone with average indoor levels approaching 4.2 pCi/L (North Carolina State Radon Office).
- Radon levels averaging over 4.0 pCi/L are considered unsafe. True or False? True. Radon in the air is measured in "picocuries per liter of air", or "pCi/L." Levels less than 4 pCi/L are considered safe, although if you can reduce the level further it's a very good idea.
- If you test your home for radon and find levels exceeding what is considered safe, you will have to move out and have the building remediated. True or False? False! This is the good news - if you need to reduce the levels of radon in your home, it is not overly expensive to do. A vent and fan system is usually the first line of defense, and will lower the radon to acceptable levels over 75% of the time.

How did you do? Learn more by going to the Environmental Protection Agency's web site (www.epa.gov) and www.radon.com.

20

Winterizing Your Home

Leaving for the Winter? Keep Your Home Safe Part 1 of 2

Another beautiful summer in the mountains is drawing to a close. It seems as if time goes faster in such a beautiful spot; or perhaps it is a function of adulthood. Either way, we are faced with the change of seasons, and the resulting chores around the house.

I've been asked numerous times about the best way to winterize your home if you leave it empty over the winter. Let's discuss this, and in another column we'll talk about what to do if you're only leaving for a few weeks.

If you go away for the entire winter, take steps to ensure you will return to a home that is clean and dry. Moisture, in either vapor or liquid state, can do very expensive damage to your home. Some of us have heard horror stories about a line bursting in the wintertime and icicles forming on the sides of the house where water was exiting openings and froze. While preventing this problem may be as simple as turning off the water main when you leave, the damage that high humidity can cause during a vacancy can be just as damaging over time.

Either follow these directions yourself, or call a qualified handyperson service or plumber. The money you spend on a professional winterization will pale compared to the damage caused by forgetting to drain a few of the pipes and then spending serious money on repairs, not to mention the emotional cost.

While some folks think it's ok to leave water in the lines and leave the heat on, I have seen instances where the heat source or the electricity failed for a long enough period of time to freeze and crack the lines. Your safest course of action is to drain the water lines. First, turn off the water at the water main. This valve is outside your home. If you do not have a "main", or you are on a well, shut the water off at the well or at the main shutoff in your basement. All of the water lines up to this shutoff point should be insulated. You do not want the pipe coming in to your home and up to your shutoff valve to freeze and crack.

Next, turn off the power to your water heater. Drain the water heater and any holding tanks. Make up large labels to place on the breaker panel and on the equipment so that others will know not to use or turn on. If you have a hot water heating system, you should consult a professional about draining this system. It may or may not be necessary depending upon the configuration.

Go to the lowest fixture in the home and begin opening faucets and draining everything. Go slowly through the house, opening up all faucets, flushing all toilets, and leaving valves open so that water can completely drain. If you can apply some positive air pressure from an air pump to water traps, it's a good idea and adds some insurance to the process. The idea is that you don't want ANY water ANY WHERE in the system.

Stay tuned! We'll finish our winterization process in the next column. For a checklist, simply go online and do a search for "Winterizing checklists."

Leaving for the Winter? Keep Your Home Safe Part 2 of 2

In the last column we talked about the procedure to winterize your mountain home if you leave it for extended periods of time. Turning off the water and draining all the fixtures is the safest way to return to a home that is clean and dry. In the last column we drained all the fixtures, leaving the faucets and valves open. We put notes on everything!

Next, open all of the outside hose bibs and leave them open. Do not leave any hoses or distribution valves attached to the bibs. Now go back inside and lift up any flexible spray hoses in showers and sinks (aha! Forgot about those, didn't you?) and drain the water from them.

Purchase some non-toxic antifreeze solution from a marine supply store or RV store. Pour some of this solution into your dishwasher and run a partial cycle to get it into the pump. Do the same with your washing machine. If this part of the procedure worries you (you do want to get enough of the antifreeze into the pumps but you do not want to run the pumps dry either) then consider getting a professional to help you.

Next, put some antifreeze solution (the non-toxic variety) into all toilet bowls, sinks, tubs, showers, and drains.

Put your large DO NOT USE – WINTERIZED labels everywhere and in several front windows of the home.

Next, close all fireplace dampers and vents. If you have a crawlspace, consider closing these vents, although there is some disagreement on the part of professionals about whether this continues to be a good idea or not. If your crawlspace is dry now and the vents are open, leave the vents open. Make sure all of the screening at the vents is intact, as you do not want any animals taking over your cozy crawlspace during the winter.

Turn off non-essential circuit breakers. Keep some lighting on, using several variable timers (available at Radio Shack or home supply stores), inside the house. Empty and clean your refrigerator. Turn off your gas supply at the tank outside.

Install a dehumidifier in the lowest part of the home and make sure the drain runs safely to a drain inside that is below the frost line so the condensate will not freeze and back up. Set the dehumidifier to about 50% and make sure you test it before leaving. You can purchase a simple hygrometer to measure humidity in most home supply stores.

Take any important papers or sentimental items and store them in a safe place in the home, or separately in a safe deposit box, or consider taking them with you.

Finally, notify your local police and fire department before you leave.

While this sounds like a lot to do, if you perform these steps or have a professional help you perform these steps, the chances of you returning to a home that is safe, dry, and intact are high and your stress will be low. No need to worry about the house.

21

Saving Energy and Money

Beat the Heat with These Energy Saving Ideas

Since this month is July, you probably have your air conditioner running full blast, don't you? Whether your air conditioner is in your window or is a central unit, you know that this time of year you are paying a lot of money to feed this beast! Consider these five money saving tips.

1. Check the filters. If they are dirty, clean or replace. Make sure the filter is the right size! You'd be amazed at how often I see filters not covering the opening, or jammed into the opening of the air return.
2. If you do not already have a programmable thermostat, then consider getting one and programming it. If you're not the technical type, read the step by step instructions in the manual or find a teenager. Once done, this "set it and forget it" system will save you hundreds of dollars over the course of a year. Before programming, figure out your schedule. What rooms will you be in the most and when? For example, put the master bedroom on 84 when you're not there, and 74 for sleeping.

3. Close the drapes or lower the shades on the sunny side of the house during the hottest portions of the day. If your windows are not tinted, consider it. Turn on fans. The amount of electricity fans draw is very small and the breeze will trick your body into thinking it's 5 degrees cooler. Fans are what they call 100% duty cycle - meaning you can leave them on all the time.
4. More than 25% of the heat in your house is absorbed through the ceiling (roof). If you don't already have an attic fan or ridge vent, consider installing one. Whereas you want ventilation in the attic, you don't want ventilation (or air movement) around doors and windows where the cool air can escape, so make sure places where utility lines enter the home are caulked and sealed and doors and windows have weather-stripping.
5. Remember what Mom said when you were a kid? She was right! Turn lights off when you leave the room. If you have track lighting, those incandescent bulbs generate enormous amounts of heat. Change over to CFLs (compact fluorescents) - you can even get CFLs that will work with dimmers. These bulbs will cost more to buy new than the incandescents, but will last longer and be much more efficient. If you keep forgetting to turn off lights, install a motion sensor that will turn them off when you leave the room.

While there are dozens of other cost savings tips for summer savings, following these five will cut at least 15% off your monthly bill.

Do You Have Energy Vampires?

Now that we have successfully navigated Halloween, do you have vampires roaming through your home? You might think they have been put back in the closet with the kid's trick or treat costumes, but beware: the vampires remain at large in your house, and it is very difficult to get rid of them. These vampires are sucking about 8% of your energy over a year's time.

I am talking about all those glowing red, green, and blue LEDs on your computer monitor, your network router, the laser beam on your garage door opener, the soft lights below your television monitor, the clocks on your microwave, the lamp at the end of the surge suppressor extension cord, and all the assorted glowing devices in your bedroom that you see when you can't sleep and wonder who turned the lights on.

These standby electrical states in our modern electronics are called "energy vampires." We tend to not think about them because they don't make noise and sometimes don't even glow. Pretty tricky, these vampires. Test your "vampire" knowledge with this short quiz.

True or False: Your telephone battery charger draws electricity when your phone is not plugged into it. True, but it's very small, less than .2 watts. The problem is that we have so many of these tiny electrical draws, they add up.

Which devices draw the <u>most</u> electricity in standby mode? A. Cordless phone; B. Alarm clock; C. The microwave; D. The electric garage door opener. Answer: D., the garage door opener. It takes a relatively large amount of standby power to continuously run the beam across your garage threshold, and to power the remote.

What item should you NOT turn off when you're not using it? A. Your computer; B. Your microwave; C. Your laser printer; D. Your toothbrush charger. Answer: C., your laser printer. These devices draw very low amounts of standby power, but will draw large amounts of power when you turn them back on after a complete shutdown.

What household device uses the <u>least</u> amount of standby power? A. Alarm Clock; B. Computer Speakers; C. Internet Modem; D. 46 inch flat screen television. Answer: D., the flat screen TV!

How many "vampires" does the typical American household have plugged into the electrical system at any given time? A. 2 B. 10 C. 15 D. 20 Answer: D., the typical household has at least 20 devices plugged in and on standby at any given time.

Is there any way to reduce this consumption? Yes. Buy an electrical extension cord that has multiple outlets on the end with an on/off switch. Use this to plug in your computer speakers, router, or pc cable modem and switch off when you're not using them.

There are some pieces of electronics that you don't want to unplug, including items that are difficult to unplug, items that respond to a remote control, and receivers for television DVRs that have to recycle (reboot) if the power is disconnected. In these cases, I would spend the $35 a year to keep these plugged in. Being aware of the other electrical wasters and unplugging them when you're not using them should save you enough money to take your family out to dinner at least several times during the year. Not a bad reward for slaying the vampires.

Home Energy Tips and Tricks

Readers say they are excited about the energy savings part of home automation, but not the complications of automation. We know that our electronics, appliances, and lighting consume a large portion of the electricity we pay for every month. Are there things we can do to lower what we spend but not over complicate our lives to do it?

The good news is yes. I am going to share tips and tricks for combining a little bit of home automation with lowering your electric bill.

The following tips are based on your lighting and hot water heating consuming at least 20% of your electric bill, and your forced air heating and cooling system consuming 30%. Here is what you can do:

Occupancy sensors. Replacing an existing light switch, these special switches (about $20) are easy to install. The sensor can tell when you are entering a room, and they will automatically turn the lights off after you leave. Do your kids ever leave the lights on? Once you begin using these, you won't know how you got along without them. Coming in to your home at night, entering the garage or carport at night, walking into the laundry room with your hands full – you'll appreciate this automation. If you're not electrical wiring experienced (always turn the breaker OFF before doing any change-outs) then an electrician can switch these out very quickly for you. Good room candidates include closets, hallways, garages, and entranceways.

Timers. Digital or analog timers can control lights and appliances to save you money. Great applications for timers include on your water heater (this is an electrician's job), on outside lights, and inside lights when you are leaving for a few days. The simple old analog timers will come back on after a power failure and not lose their settings (albeit a few hours later if the outage has been substantial). You can also use these timers to turn off your standby electronics (called "vampire power") when you're not using them. It will not hurt these devices to be turned off for a portion of the day. Simply plug them into a surge protected outlet strip, then plug the strip into a timer and plug the timer into the wall. Devices to plug into this strip include all varieties of chargers (except, of course, if you are actually charging something), instant on TVs, the audio speakers for your computer . . . you get the idea.

Programmable thermostat for your heat pump/gas furnace. Programming when you want the air conditioner to come on and telling it when to go off should cut another 5-10% from your electric bill. If you're unsure about what to get, and how to program it, just find one of your tech friends to help, or you can get a pro to install it. Honeywell makes a unit for $119 that uses your computer to program it and is adjustable through your Wi-Fi network. Another unit, called the "Nest" learns the program from how you adjust the temperatures over a few days. You never have to actually figure out a program for it.

15% or more savings on your electricity bill – not bad for installing some easy automation.

Take the Electricity and Energy Quiz

A reader writes: "Please give me more information on saving money by reducing electricity consumption. I know that the little stuff adds up, but what other things don't I know? The quiz format makes it interesting."

The average American household spends nearly $2,000 a year on electricity. If we can knock this down by 10-15%, that's a healthy chunk of money. Take the following quiz and I'll explain the answers.

1. Some of the biggest gains in energy savings in our homes has been from improvements in: A. Window glass B. Roofing materials C. Televisions D. Appliances. If you answered D., Appliances, you're correct. Modern refrigerators, for example, use about 70% less electricity than the ones made in 1975. Clothes washers and dryers are big consumers of energy, and have become much more efficient over the last 15 years. If you have a very old set of appliances that are still working fine, you are paying about twice as much for electricity to run them. Maybe this will make you feel better when your old Maytags break and you finally have to go buy new ones.
2. One third of the average household's electric bill is spent on: A. Clothes washing and drying B. The oven C. The televisions D. Air conditioning and water heater. If you answered D., you are correct. Electric air conditioning and electric water heaters are big energy eaters – up to 50% of your electric bill.
3. Which of the following home improvements will make the biggest contribution to reducing the energy consumption in your home: A. Painting the house a light color B. Sealing air leaks, including leaky ducts C. Installing new triple glazed windows D. Adding insulation to the attic. Did you answer B., sealing air leaks? Great! In order of savings, sealing leaks, including in ductwork, is number one, adding insulation is second, and upgrading windows is third. The color of your house is

probably less important than your roof color, which can make a difference.

4. You can reduce your lighting electric consumption by 20% by: A. Installing fluorescent strip lights B. Replacing your incandescent bulbs with compact florescent light bulbs ("CFLs") C. Turning off the lights when you leave the room D. Both B. and C. If you answered D., you are correct. CFL bulbs use 70% less energy than incandescent and last ten times longer. Replacing just five of your high wattage incandescents can save you $75 or more a year. The new LED bulbs are great energy savers as well, but they are still expensive to buy.
5. What is true about ceiling fans? A. You should only use the fans in summer. B. Ceiling fans use extra energy, you shouldn't use them at all. C. Turning them off and on will use more electricity so leave them on. D. Ceiling fans work well in both summer and winter to improve comfort. Answer D. is correct. In winter, run them clockwise (upwards) to spread the heat that rises, and counterclockwise in the summer.

How did you do? If you got 3 or more right you're doing better than most households, and you're probably saving quite a bit on your energy consumption. Every little bit counts; be proud of yourself.

Feeling a Draft?

Winter has arrived in the mountains and along with it, those typical freezing temperatures at night. Are you warm and snug in your home as the wind howls outside, or do you sometimes feel a draft of cool air? Air leaks in your home will reveal themselves readily at this time of year. Energy bills go up, and we wonder how we can stay toasty and still afford to pay for it.

If you really want to save money on energy costs, the wintertime is the best time to determine if and where your homes has air leaks. If you seal these up, you'll see the difference in your bills almost immediately. These "air flow" savings will carry into the summer air conditioning months as well.

Most of us think about the leaks around windows and doors. While these certainly contribute to the overall leakage, the significant leaks are actually in the attic, walls, and basement. You should still "audit" the seals around windows and doors, since these are easy fixes – if you see daylight anywhere around the frames of windows or doors you should install new or improved gaskets and seals. This is usually a do it yourself job and involves some common sense and a trip to a hardware or home improvement store. Folks ask me if they should improve or update their windows to save on energy costs. My answer is usually no. This is because most of the time the cost to replace your windows is higher than the energy loss (cost) over time. Are your windows cracked, rotted, or split? Then replacement may be a good idea. If they are in good condition but just not double glazed, you can repair them and also add storm windows to save heat loss. If you are not sure what to do, have a qualified contractor or a green energy professional take a look at them and advise you.

Where you'll save the most money is by searching out the "holes" that the original builder forgot to seal. As heated air flows through these areas, it flows upwards into your attic, drawing cold air in at the lower levels. You can tackle this search and seal job yourself, or you can hire a qualified contractor.

Attic hatch. Whether this is a square hatch in the ceiling or pull down stairs, there's a good chance that there is zero insulation between it and your attic. The best solution is a piece of rigid foam with Velcro that can be easily removed for access.

Wiring holes. Turn on your home's exhaust fans and hold a lighted incense stick near your electrical outlets. If the smoke blows sideways, then you have air entering or exiting the walls. Expandable foam works well in these locations.

Vents that go through your walls. Check plumbing pipes and stacks – under sinks, in the utility room, in the attic, in the laundry room - do you feel any cool moving air or are there any holes? Use caulking or foam in these areas.

If you want to be extra diligent, you should also check and seal behind knee walls in attics, around recessed lights (buy special insulation cans for these), spaces under built in furniture and appliances, furnace ducts, and any other areas where you feel cool air coming in to the house. Count your savings and stay warm!

Readers Ask About Energy Recovery Ventilation

A few weeks ago, I mentioned an "ERV" or Energy Recovery Ventilation system could help keep your home air from getting stale and keep the indoor pollution levels lower by exiting air from the home and bringing in fresh air. Several readers wrote to me and asked me to explain what this device is and how it works.

The last 20 years of building technology have brought much tighter homes – a move to 2X6 exterior walls making more room for insulation, window glass with higher insulation levels, better weather stripping, and much more efficient heating and cooling systems. But newer homes without an ERV may trap contaminants indoors.

So, the ERV, or "HRV" if the system is heat recovery only, is typically added to the heating and cooling system for tightly built homes to provide air exchange and reduce the amount of indoor pollution. The ERV consists of a large box with fans, filters, and electronics. It's hooked in to your central heat and air ductwork in such a way that it can exhaust stale air outdoors and bring in an equal quantity of fresh air to replace it while also capturing the heat (or cool) from the outgoing air to warm up (or cool) the incoming air. An ERV can also capture some of the humidity in winter and return it to the house, and exhaust some of the excess humidity in the summer.

Older, less air-tight homes may cost more to heat and cool, but the higher exchange of air make them fairly healthy in terms of the air. Not so with the much tighter new homes, which can have indoor pollution levels that are 2-5 times higher than the outside air. New materials inside a new home – from carpet to vinyl to furniture – take months and even years to off-gas formaldehyde, flame retardant chemicals, and other materials used in the manufacturing process.

Can you just open some windows to clear out stale air? Absolutely you can. There's something wonderful about opening the windows in the spring and inhaling the mountain fresh air. But, what if allergy season has arrived and you're trying to hide from the pollen? What if its winter and opening the windows is impractical? In these cases, indoor air can degrade significantly, even with heating and cooling systems running, causing scratchy throat, coughing, and watery eyes.

Now that you know all about an energy recovery ventilation system, you are probably thinking that all new homes must have them. But no, not all builders install them, and the building code does not require them. The code assumes that occupants can simply open the windows for ventilation. But as we just observed, this can be impractical.

Can an ERV be added to an existing central heating and air conditioning system? It can indeed. The installers might get cranky if you have a small HVAC closet, but they will get creative. You might have to give up closet space. They will install 2 vents to the outside – an intake and an exhaust. Filters will clean incoming air, and you can balance the system with a control that allows you to choose how many minutes an hour you want the system to operate.

Now you know!

Choices in Programmable Thermostats

Do you know where your energy dollar is going? Do you assume that there is little that you can do about your electric and propane bills? Did you know that heating and cooling your home represents at least 40% of the bill? Most experts tracking this sort of data will tell you that this adds up to an average of $1,800 a year. What if you could save 10% of this money, or $180 a year?

Enter programmable thermostats. I can hear you now: "Programmable thermostats? I can't even get my digital clock set. Forget it." Programmable thermostats used to be complicated. I remember copying a page out of the user manual and putting it in the kitchen drawer for every time I had to change the batteries in the unit. But I took the time to figure it out because I knew it would make a difference in my budget. So I can understand your reaction. But times have changed, and I encourage you to reconsider.

Here are some reasons why you should consider a new programmable thermostat.

- Easier to use. The better units have solved the "difficult to use" complaints, with intuitive interfaces and great screens.
- It won't forget. If you are setting your central air conditioner to 80°F when you leave the house in the morning, how many times do you forget, leaving the house at 72° all day, running up your electric bill?
- Cool features. Newer thermostats have filter change reminders, humidity reminders, live weather, and energy usage data.
- Connected. The higher end units connect to the internet via WI-FI and can be controlled remotely through your smartphone, tablet, or computer. Install one in your second home and know quickly if there is a problem.

How do you know which thermostat to buy? Will it work with your heating and air conditioning system? Are the new thermostats expensive?

To decide what to buy, I recommend you do some research. Read the reviews on Amazon.com. You should decide whether you are a gadget lover (like me) who likes to fool with things, or a person who does not really want to know the details, but just wants it to work.

If you are the gadget geek, then spring for one of the top of the line WI-FI models. The Venstar T5800 ($165) and ecobee EB-STAT-02 ($274) are top rated for features and ease of use by Consumer Reports. Also highly rated are the Nest Learning Thermostat ($249) and the Honeywell Prestige HD ($250).

If you want something cheaper, then I recommend the Lux Smart Temp 9600 and the Honeywell Focus Pro (both about $70). Unfortunately, spending much less than this will get you back to "hard to program" category, and you may never use it.

How will you know if the unit you want will work with your particular heating and cooling system? Find out whether your system is straight heat pump, or if your furnace also uses propane (dual fuel). Go down to your favorite home improvement store with this info and the manufacturer name and they will be able to advise you.

The best trick of all is to talk one of your children or your neighbor's child into programming the thermostat. Children intuitively understand these things, while some of us from an earlier time find programming gadgets to be like solving differential equations.

22

Water, Plumbing, and Wells

Your Healthy Home: Water Leaks

Water leaks are the bane of every home owner. I have heard true life horror stories of people leaving their home to go on a cruise or vacation only to get a call from a neighbor who says that water is running out their front door. I saw icicles stretching down from the second story siding where the pipes froze and burst inside a cabin retreat once when I arrived for an inspection!

Are you water aware? What do you do when you leave your home for an extended period of time? Most of us, if we are leaving for many days, turn the water off where it enters the home, or sometimes outside or in the crawlspace at the well pump. But when you think about it, we could leave for work and have a burst or leaking pipe do the same amount of damage as leaving for days at a time.

No, I'm not trying to give you nightmares, although the scenarios above are pretty scary. Here are some tips for making sure the insurance company does not get a call from you anytime soon about flooding. Even if you do not have water that "escapes" the system on to your floors, you can still have leaks internally in appliances that run your water bill up, or consume running time on your well pump.

For example, a toilet that leaks internally (flapper not seated) can waste over 200 gallons of water a day.

First, know where your water shutoff is. Make sure everyone in the house knows where it is. This way, in an emergency, you can cut it off quickly. Second, take a look at all of your water gulping appliances. Are there shutoffs nearby? Most appliances - especially toilets and faucets - have their own shutoff valve so that maintenance is possible. These valves themselves can leak.

Second, take the time every couple of months to look under sinks for seepage. Check the shower head where it meets the arm coming out of the wall (as a home inspector, I find leaks here all the time - if left unchecked, these will grow and spray water above the shower enclosure), look behind your laundry appliances, the bottom of your refrigerator and dishwasher, and around the faucets in your shower and sinks - sometimes these will seep from the base of the fixture. Water dripping from your shower faucets can seep behind the fixture and damage walls - if you see any dripping from the faucet handle, you should have it repaired. Take a good look at where the water is running during a shower - if there is any possibility that water can run into cracks in tile or behind the tub faucet, caulk these areas. Check tile thoroughly - small cracks can develop over time and allow water to run in under the shower pan.

Third, check your toilets. Toilets can leak without making noise - but first take the toilet tank cover off and take a good listen. 90% of the time, if you have a leak in the toilet internals, you will hear it. It might be a slight hissing, or just the sound of intermittent drips. Non-leaking toilets will be completely silent. If you are unsure, put a few drops of food coloring in the tank (not the bowl). If the color shows up in the bowl, your flapper valve at the bottom of the tank is leaking. These are easy to replace and cost just a few dollars.

Your Healthy Home: Water Leaks 2

Last week we talked about water leaks in your home. Let's continue our discussion by talking about steps you can take to minimize the chance of water damage, and how you can conserve water. Even if you have a spring or well, the water supply is not endless, and lots of water usage or hidden leaks can put a strain on your water pump equipment.

Even brand new homes can develop a water leak. One new home that we had finished up about 2 months earlier is an example - the homeowner decided to run the water pressure up another 10 PSI. This resulted in several leaks from faucet pipes under sinks - the plumbing contractor had not tightened these completely and it didn't show up until the higher pressure was applied. When inspecting new construction, I take considerable time to really test all of the fixtures for this reason.

Last week we talked about what you can do to inspect your home for leaks and seepages, and that some of these can be hidden. This is particularly true if your home has a crawlspace. While from an inspector's point of view a crawlspace is wonderful because we can SEE everything that is going on, from a homeowner's point of view crawlspaces are not that fun! They are hard to get into and inevitably have an assortment of creatures from spiders to mice to deal with as we move on our hands and knees towards the pump, water heater, or air handler. I still have not figured out why a builder really believes you'd rather trade space for a water heater for a clothes closet if you have to go into a dark crawlspace to service your water heater! At any rate, what the crawlspace is good for is looking for plumbing leaks! So, in addition to the other checklist items I mentioned, go into your crawlspace if you have one, or find an adventurous teenager to go in there and LOOK and LISTEN for drips. Unless you do this, you will never know about these leaks until the next tradesperson goes in there.

Is there anything you can do proactively, besides regular inspections, to stave off a day when water gets out of piping and damages something in your home? You would think that in our day and age of modern devices and electronics, that someone would think of this. Well they have - and in a variety of forms.

These alarm devices come in two varieties. One is a passive system that sits on the floor near the fixture, such as behind your toilet or under the refrigerator, and sets off an alarm if it senses moisture. These are inexpensive and easy to install. They can be battery operated for simplicity. The problem with these is that they depend upon someone hearing the alarm and turning the water off. What if you are not home?

A better system, but more expensive, is the active detection system. This device will detect a leak AND actually shut off the water flow to the appliance. This system needs to be installed by a professional, since it involves putting a valve in-line with the appliance. Better still is a whole house active leak and shutoff system. This works by having either a wired or wireless sensor at every appliance, communicating with a central brain connected to a water shutoff valve for your entire home. This is installed much like a home security system, with a connection to an alarm system if you have one.

These systems can get very sophisticated and there are numerous products to choose from. You should consult a qualified plumber for more information if you decide to install these products; some home alarm installers also provide these systems.

Next week we will complete our water discussion with easy tips for water conservation.

Your Healthy Home: Water Leaks 3

Last week we talked about water leaks in your home and things you can do to prevent them or find out about them fast enough to do something about it before there is extensive damage. We talked about leak warning systems. If you are constructing a new home, be sure to integrate a leak detection system into the construction process. It is easy and not expensive. However, most of us are not building new homes right now, and need to address the water leak problem with easy solutions. My advice is to purchase some of the inexpensive leak detectors that are battery operated and sit on the floor behind items that might spring leaks: water heater pan, condensate pan for your air conditioner, behind toilets right next to the junction of the floor and toilet base at the rear, under the kitchen sink, etc. You can go to a home improvement store for these or they are available on popular online buying sites, such as AMAZON. Simply type in "water leak detectors" and you'll have many pages of results.

Back up this solution with twice a year visual inspections of your plumbing. When was the last time you stuck your head inside the kitchen sink cabinet and looked up? I am guessing never! Well, if you do, you will catch areas that are about to spring a leak, such as a rusted drain to sink connection, or a pinhole rust stain on your disposal just waiting to spray water under your cabinet. This will not take long but will pay off in heading off problems – just take a good look at all your pipes – in bathrooms, kitchens, laundry room, and utility areas and any other place you have access to where there are plumbing pipes. Don't get confused with the propane piping – these look like they could be carrying water but of course do not, and these gas pipes are sometimes black (iron) metal. Most recent construction (in the last 20 years) will use CPVC piping, or Chlorinated Poly Vinyl Chloride, for the potable water supply, both hot and cold. CPVC exhibits excellent insulation and durability characteristics, making it an improvement over previous materials. You can recognize CPVC piping by its beige color (regular PVC is white and is used for waste piping and incoming water supply).

This discussion of water leaks would not be complete without a mention of Polybutylene water pipes. These pipes began being used in the 1970's and were considered an advancement over previous materials. From 1978 to as recently as 1995, millions of homes were outfitted with "PB" pipe. But in the 1980's, less than 10 years after the installations began, these pipes started leaking, and in catastrophic ways – from multiple burst pipes, to leaking joints. The PB material deteriorates from the inside out, so it's impossible to inspect. Failures may occur in systems with plastic fittings, metal fittings, and manifold-type systems that look fine even to the trained eye. I know you're asking this question: "How do I tell if I have this type of piping?" The answer is straightforward but also a little complicated. Generally, Polybutylene pipes are grey in color, and have plastic or copper fittings at joints. But there are other materials that are not PB and can fool you. Look under your sink where the pipes come in at the bottom of the cabinet (don't confuse these supply lines with the lines that run from the valve to the faucet, because grey color here is ok.) If you do have grey water supply pipes in your home, you should have a plumber look at it to determine what it is. Then you will know for sure. If you do have Polybutylene piping, follow the plumber's recommendation on inspections. The joint materials improved in the 1990's, and the material in your home, if you have it, may not be a problem. If you want to learn more about PB pipes, simply type "PB pipes" into your internet browser and you will probably see more than you wanted to know!

Should You Buy a Water Leak Detector?

A water leak in your home can be a nightmare, depending on its location and severity. Last week we talked about what you can do to inspect your home for leaks and seepages, and that some of these can be hidden. This is particularly true if your home has a crawlspace. While from an inspector's point of view a crawlspace is wonderful because we can SEE everything that is going on, from a homeowner's point of view crawlspaces are not fun places. They are hard to get into and inevitably have an assortment of creatures to deal with as we move on our hands and knees towards the pump, water heater, or air handler. I still have not figured out why a builder really believes you'd rather trade the space for a water heater for a storage closet if you have to go into a dark crawlspace to service your water heater or turn the water on and off. The crawlspace should be inspected for water leaks from time to time because otherwise the leaks will go unnoticed and you'll wonder why your water bill went up. So, find an adventurous teenager to go in there and LOOK and LISTEN for drips.

Are you thinking there must be a better way to identify leaks? There is – water leak alarms!

One water leak alarm type is a passive system that sits on the floor near the fixture, such as behind your toilet or under the refrigerator, and sets off an alarm if it senses moisture. These are very inexpensive and easy to install. The problem with these is that they depend upon someone hearing the alarm and turning the water off. What if you are not home? And the battery needs to be replaced periodically. But they are better than having nothing.

A better system, but more expensive, is the active detection system. This device will detect a leak AND actually shut off the water flow to the appliance. This system needs to be installed by a professional, since it involves putting a valve in-line with the appliance.

Better still is a whole house active leak and shutoff system. This works by having either a wired or wireless sensor at every appliance, communicating with an electronics board connected to a water shutoff valve for your entire home. This is installed much like a home security system, with a connection to an alarm system if you have one. Typically the whole house option is done when constructing a new home, since retrofitting the system later is more expensive.

These systems can get very sophisticated and there are numerous products to choose from. You should consult a qualified plumber for more information if you decide to install these products; some home alarm installers also provide these systems.

Your Healthy Home: Water Saving Tips

"Turn out those lights!" Turn off that faucet!" "Close that door, do you think you're living in a barn?"

Have you heard any of these before? I bet you have. I certainly have. The experience of receiving as well as giving these scoldings may not have been a completely happy experience. Isn't it interesting that as kids we thought it was "nagging" on the part of our parents. Now, as adults, we say these same things to our kids/friends/spouses. Intuitively, we know that wastefulness is not good. But often we're not prepared to make the sacrifices necessary to create thrifty habits. Put any negative feelings aside from your experiences as a child – now you have the power to really put resource savings – in this case, water – to work for you in some ways that are not torturous. By following the following tips, you will save money every month on your water bill. If you have a well, as many of us do, the savings will be in the form of lower maintenance and repair costs. The following tips are my favorites; if you want to do more, there is a wealth of advice on the internet under "water saving tips".

The bathroom. 75% of water used indoors is used in the bathroom. This is no surprise, but it really adds up: 7,000 gallons is the monthly average if there are 4 people in your house and you have average flow fixtures. Just one 10 minute shower can take 50 or more gallons! No, don't stop taking showers, but read on.

I know you hate low flow showerheads, so I won't talk about it. So, to save some water, how about tightening up your shower routine? If you shave a couple of minutes off your total time, you'll save an average of 7,000 gallons a year. And if you're really serious, turn off the water while you're soaping up.

The toilet takes 25% of the water used in the bathroom. If your toilet is the old kind, and uses 5 gallons of water per flush, you can place specially designed "tank banks" or even a plastic jug with pebbles in your toilet tank to reduce the amount of water that gets flushed. See how little water will work – but don't reduce it so much that everyone starts flushing twice. If you're remodeling, replace the toilet with a modern unit, which uses less than 2 gallons per flush.

Leaks. Make sure your toilets are not leaking. Put some color dye in the tank and see if it migrates to the bowl. If you are on a municipal water supply, go look at the meter when all your water fixtures are off. Is it moving? If so, you have a leak somewhere.

At the sink. Don't run the water unless there's something happening under the faucet! This means turn off the water when the toothbrush is not under the water flow, and do the same thing while shaving. Savings: 3,000 gallons a year.

Drips. Search out and fix those little faucet drips! As much as you think they're no big deal, they really are: one small drip per second adds up to 6 or more gallons of water a day (2,000 gallons a year for one drip). Multiple drips . . . you get the picture!

In addition to saving money, being water conservation conscious will make you feel as if you're doing your part in the grand scheme of helping the planet. And you are!

Your Healthy Home: Water Saving Tips 2

Last week we talked about conserving water in your home, and that 75% of the water that you use is in the bathroom. What about the rest of the house? You guessed it: the kitchen is the second highest water waster.

To save water anywhere in our home takes a conscious effort to change our current habits. Even during the drought, there were things we were doing in our daily routine that could have saved water and money if we'd known how and made the decision to re-think our daily routine. Although the following tips apply to the kitchen, you can take these same ideas and apply them across the board to everything else you do in the home. If you have children, start a game with them to see how much water you can save. For example, every time someone leaves a faucet running, tell them that the first person to point the wasteful action out gets 10 points toward some reward. You'll be amazed at what they come up with!

In the kitchen, the water tap is the biggest water waster. We tend to let it run without doing anything under the faucet, or we're running it waiting for hot water to arrive. A trick is to stopper the sink when you're rinsing plates, either when washing them, or when rinsing them for the dishwasher, and then use the water in the sink for washing pots and pans. If you do have a dishwasher, you should actually avoid pre-rinsing the dishes, as dishwashers are built to handle that crud on the plates; just scrape the leftovers into the garbage (dog dish?). If you have a disposal, use it sparingly and only for the light debris that is left in the sink; using the disposal uses a lot of water, and if you have a septic system, using the disposal for heavy items such as vegetable matter will slow down the all-important septic breakdown process.

If your dishwasher has a short cycle, try using it for full loads too. The magazine Consumer Reports has found that dishwashers do a remarkably good job of cleaning, even when the plates and pans have debris on them. If the short cycle works well for the full loads, it will save water.

Make sure your kitchen faucet isn't leaking. Because of the heavy use they typically get, single handle faucets can begin seeping from the base. To test this, move the handle all around in a complete circle and take a good look at the area around the bottom of the handle. This seeping can go unnoticed and waste a lot of water in a short period of time.

Do you have to run the tap a long time to get hot water? If so, this is wasted water unless you capture it somehow. There is an ongoing debate on the part of home builders as well as energy consultants about tankless, or "instant" water heaters versus conventional tank water heaters outfitted with recirculating pumps which deliver instant hot water to the tap. Which system saves more money and is more efficient? We'll tackle this in a future article.

Your Healthy Home: Instant Hot Water? Part 1

In the summer, you might not mind jumping into a shower with out of the tap temperatures, but here in our mountains in wintertime this is out of the question unless you're Chuck Norris. One of the worst things that can happen in your house is to hear, "No hot water!" discovering that the water heater has failed just as you're about to step into the shower on a chilly morning.

Since we've been discussing how to save energy and money, and not waste water, is there a way to have your cake and eat it too, or have instant hot water at your water tap and not spend a fortune doing it? As soon as you turn on the faucet, hot water is there – no running fixtures for minutes waiting for the hot water to show up. This is an interesting debate, as heating hot water in our homes represents 30% of our total energy bill.

If you do want instant hot water, you have some choices. The first is tankless, or "instant" water heaters that heat water quickly and deliver it to your tap on demand. Or you can use a conventional tank type water heater outfitted with a recirculating pump to deliver instant hot water. Which system saves more money and is more efficient? Which one works better? Should you consider a hot water circulator? When your hot water heater fails should you toss it and install a tankless water heater?

The answers to these questions will depend on how big your house is, what your hot water demand is, and whether you're running only electricity or a combination of electricity and gas (propane). You'll want to consult a qualified contractor if you decide to go with any of these instant hot water solutions to make sure they're matched to your needs. And remember that there is an ongoing debate on the part of contractors about which system is best! Let's examine some of the pros and cons of each solution.

Tankless "instant" hot water heaters – these water heaters are small and conveniently fit on a wall or under a cabinet. They supply hot water on demand and do not operate when you're not calling for hot water. Because of this, they are very efficient, and can reduce your heating bill by 25%. These units are reliable, and don't "run out" of hot water, since they use high powered propane gas burners and a heat exchanger to heat the water as it flows through the unit. The units are durable and may last for over 20 years.

Before you run out and buy one, though, consider some of the drawbacks. Even these "instant" heaters will not supply "instant" hot water if your home is large and spread out, and there's a lot of piping. In this case you might want to consider more than one unit. These heaters are also more expensive compared to conventional tanks, costing about twice to three times as much. Considering that 25% savings might represent $75 a year and the unit costs $1,200 instead of $350 for a conventional tank unit, it will take a while to reach a break-even point. Tankless units require careful venting because they are propane fired. You can install electricity only units, but these are small and usually cannot supply enough hot water on demand for an average sized house by themselves.

Tankless water heaters can also be inconsistent in delivering hot water. The cold water must be pushed through the system before being delivered to your showerhead. And tankless units require annual maintenance – just like the tank version – to keep scale and buildup out of the works. Lastly, tankless water heater retrofitting in an already built home can be expensive because of the required electronics, piping, and venting.

Your Healthy Home: Instant Hot Water? Part 2

Last week we talked about how to have our shower and heat it too – or how to have instant hot water on demand and not spend a fortune.

Is this possible? We talked about tankless water heaters – very efficient and durable, but expensive to purchase and retrofit.

Conventional water heaters – the tanks you are most familiar with – are not expensive to purchase or install, and last 12-18 years or more with maintenance. You can install an electric version if your home does not have a propane gas supply. And there are plenty of models to choose from, including newer, more efficient versions. In general, electric water heaters are cheaper to buy, easier to install (no gas pipes or venting needed), and safer (no risk of gas leaks or venting combustion problems), and are more efficient than gas fired tanks. And with gas prices (propane) going higher, the cost savings of using gas is not as dramatic – which was the only good reason to use a gas water heater in the first place.

The biggest drawback for conventional water heaters – gas or electric - is that you're keeping a big tank – 40 or 60 gallons – hot all the time waiting for the moment when you need it. You can imagine how wasteful this is. If you add a hot water circulator to this setup, now you're running hot water continuously through all of your hot water pipes, just waiting for you to turn on the tap. This situation is wonderful from a comfort and convenience standpoint, but it's expensive.

I said earlier that the most efficient solution will depend on a lot of factors in your home and there is no ready answer. But we can definitely come up with some conclusions here that should help you have the best of all possible worlds when it comes to your water heating bill and your convenience level.

If you currently have a standard 30 to 40 gallon hot water tank, as most of us do, you can save 15-30% of what you're spending right now if you have not already done the following:

Install a timer. Why heat hot water if you're away from the house? Have an electrician install a heavy-duty timer on your water heater, and set the timer so that the tank delivers hot water only when you need it. For our household, this means 6-8am and 5-7pm. If you have a recirculating pump for the hot water (this is the have your cake and eat it too part), put a timer on this too – this timer can be bought at any home improvement store. Set this timer to the same settings as the water heater timer. In my opinion, a recirculating pump operated this way will not cost much more, but will deliver instant hot water and save water because you won't have to wait for it at the tap.

Buy a blanket for your heater. Make sure this is designed for the heater. Don't just wrap batts around it; you might end up blocking vents or access areas. If your heater is already in a warm room of the house, such as next to the furnace, you can skip this step.

Check the temperature. Make sure the temperature is not set too high. The hotter your water, the more it will cost to heat it. Most households do just fine with 125 degrees – hot enough to prevent bacteria from growing in the tank, but not so hot as to be scalding.

Now you have a good working knowledge of the choices available and the information you need to save some money. Enjoy your hot shower!

Your Healthy Home: Instant Hot Water? Part 3 Loop

Last week we talked about "instant" hot water and how to have the advantages of a hot water circulation pump without spending a lot of money using your existing, conventional hot water tank with a powered circulator – but adding timers to the arrangement so that it's not continuing to circulate hot water when you do not need it.

The solutions we've talked about apply to existing construction, because most of us are not building a home right now. However, what if you are building a home, or planning one? If this is the case, there are some things you can do – whether you are an owner-builder or you've hired a contractor. In either case, you will need to be on top of things at the job, because subcontractors are not all aware of these energy efficient ideas. What we're about to talk about takes more work when the system is going in, but will pay off handsomely later in your smaller electric bill.

If you are building a home, you can specify that a "Hot Water Gravity Recirculating Loop" hot water system be put in. One Progress reader wrote to me to say he had "instant hot water without all this pump stuff" and that is exactly what this system does. Why this system is not installed in all homes with basements (or where the hot water heater is downstairs) is beyond me, except that it does take more up front plumbing work when the system is assembled.

A loop hot water system takes advantage of some free physics facts (say that fast 3 times). Since our mountain homes are mostly vertical (basements or crawlspaces, then a first floor and sometimes a second floor or a loft), this system uses the configuration beautifully. The physics facts: heat rises and cold falls. Imagine a simple plastic loop filled with cold water in your hands (the size of a hula hoop if you remember those), and you heat one side of it. The hot water will rise . . . and the cold side water will fall, and the water will then begin moving slowly in the loop without any help at all from you. Great! This is an example of convection (energy transfer) that we also see naturally occurring in the atmosphere. Let's harness it.

Conventional plumbing moves through pipes from your hot water heater to your faucets and then stops. A loop system adds a return line which connects to the bottom of your water heater. Voila! You have instant hot water any time of the day or night at all fixtures.

But wait! Think of all that hot water circulating in your home. Won't this cost extra to heat? The answer is yes, it will. We handle this by insulating all of the hot water pipes – until the end of the loop, where we still need cooler water in order to make the system work. Insulating the hot water pipes will also reduce naturally caused noise in the system, a bonus.

A gravity feed hot water recirculating loop will deliver hot water to all of your fixtures all of the time, is simple, and the extra operating cost is extremely low. If you go on trips or use your home seasonally, simply turn off your water heater (a good idea with any system) while you are gone. You can even use timers on this system to increase your savings.

If you are building a home now, or you're planning a new home in the future – and it has more than one story (has a basement), find a builder who knows how to install a hot water loop system. If you're an owner-builder, find a knowledgeable plumbing contractor who has installed these systems. They are more work to put in – and the details – insulation, standoff brackets, and positioning – are important, so you should make sure they know what you want, and you should supervise the work carefully to make sure it's done right.

Your Healthy Home: Instant Hot Water? Part 4 Solar

Over the last few weeks we've talked about the various ways to get hot water to your tap without spending a lot of money in electricity or gas to do it. The advantage of having "instant" hot water at your fixtures is not only comfort and convenience, but running the water for 30 to 60 seconds or more waiting for the hot water to show up for your shower is wasteful.

Last week we talked about gravity fed hot water loop systems you can install in your home during construction. Our mountain homes lend themselves to these installations because of the vertical nature of our buildings – we can put the hot water heater in the lowest part of the house and run the hot water loop vertically, enhancing the natural ability of hot water to rise and cool water to fall. Let's talk about another system that uses this same principle - solar water heating. Should you consider it?

A solar water heating system uses the same principles of the hot water loop (convection), but we replace the hot water heater – whether electric or gas – with a storage tank. The system can be completely passive – using no electricity at all, or it can include a pump to move the hot water as it flows through solar panels, or collectors, alongside your home or on the roof of your home. Some systems use a heat exchanger with a special fluid inside a tank to heat water and then circulate it with a pump. Solar systems also employ backup electrical heating elements in case there is more hot water demand or if there is a problem in the solar mechanics.

It would seem that a solar hot water system using both the sun and the natural principle of convection would give us nearly free hot water. If this is true, then why doesn't every home have solar hot water heating? Solar power and solar water heating have been available and used for hundreds of years, and the systems have gotten more sophisticated, so it should be a simple matter. Up to 30% of our gas or electric bill goes to heat water in our homes, so the incentive is certainly there.

The reason is just what you are thinking: up-front cost! Most builders throw in a conventional 40 or 50 gallon electric hot water heater for a total installed cost of about $150. If they installed an $8,000 solar water system they would not only need to recoup that cost and make a profit, but they would also discover that not every home buyer cares about saving money over the long term, or necessarily cares about being energy resourceful. And not everyone likes the look of the solar panels on their roof, although modern technology has made them almost invisible.

However, if you are planning to build your own home, you should definitely consider a solar hot water heating system. The pay-back period will be about 5-7 years, and the system will cost between $3,000 and $8,000 installed. And yes, you can even retrofit your existing home. These systems are very durable with most warranties running 20 to 30 years.

Your Healthy Home: Instant Hot Water? Part 5: Heat Pump Water Heaters

Over the last few weeks we've talked about the fact that heating the water in our homes takes up to 30% of our energy dollar – no small thing. We want and need convenience, and "instant" hot water is the ultimate creature comfort. We found out that it is possible to have hot water at the tap and save money at the same time, especially if we're retro-fitting or building a new structure. We can also save money by adding timers and recirculators to older equipment. Today we'll wrap up our review of hot water heating alternatives by exploring the newest technology on the market: Heat Pump Water Heaters.

Heat pump water heaters? Well this sounds confusing. I know you're thinking, "I'm not even sure what a "heat pump" is, much less a "heat pump water heater"! You may have a "heat pump" for your air conditioning – both heating and cooling – in your home now, so it sounds familiar to you, but you may not understand how it works. We'll go into that in a future article. Suffice to say that a "heat pump" combines a condenser unit (some folks call this the "compressor") which is outside, with an "air handler" with ductwork inside your house, to heat AND cool the air. During the heating season, the pump moves heat from the outdoors into your house; during the cooling season, the pump uses a reversing valve to move heat from your house into the outdoors. Because they move heat rather than generate heat, heat pumps can supply up to 4 times more energy than they consume. Heat pumps use electricity to move the heat around, not propane.

So, a "heat pump water heater" is a refrigerator in reverse. A refrigerator pulls heat from inside itself and dumps it into the surrounding room; a heat pump water heater pulls heat from the surrounding air in the room and dumps it—at a higher temperature—into a tank to heat water. Ok, making sense? The amazing and wonderful thing about these new water heaters, or "HPWHs" as they're often called, is that they are 2 to 3 times more energy efficient than a standard electric resistance heater, so they draw less than one-third the power. That's quite a cost savings.

Wait, you say - we know we can't get something for nothing, so what's the drawback of replacing our old water heater with a HPWH? You're right; there are some drawbacks, but not many. First, a heat pump water heater will cost you more to buy up front. The payback period, though, is only 2-5 years depending on the model you choose. And there are some important installation requirements, such as locating the unit in a warm area of the home and providing plenty of space around it. Remember that this unit will COOL the room that it is in because its taking HEAT out of the AIR.

So the next time you hear someone in your house shout, "No hot water!" as they're stepping into their morning shower – you have all the information you need to choose your next water heater.

Is Your Well Water Safe?

Most of us are familiar with wells here in the mountains. A few municipalities have "city water," or a public water supply, but this is the exception rather than the rule. You may have moved to this area knowing absolutely nothing about wells, and were unconcerned when you moved in about the quality of the water because it was not something you worried about before. Or, you might have gotten a home inspection before buying your home that included a water quality test. In either case, if your home has a well, there are some things you should know.

15% of the U.S. population gets their drinking water from private wells, and these wells are not regulated by any federal or state agency. This may come as a surprise to you. The EPA (Environmental Protection Agency) regulates public water supplies, but not private supplies. Well drillers follow proper procedures when drilling and installing well equipment, but then the homeowner is on their own. Over time, problems can develop with private wells that can cause serious and chronic health problems.

Well water contaminants can include bacteria, microorganisms, and parasites; radioactive elements including radon; nitrates and nitrites; and heavy metals such as arsenic and lead. A filter at your tap will not stop or filter many of these elements.

If you have your own well, here's how to stay safe.

Inspections. If you are buying a home with a private well, you absolutely should have a professional well drilling company inspect the well and test the water. This inspection will include checks of the mechanical and electrical components of your well, a site review for possible drainage or contamination issues, and a full analysis of the water quality. Doing this should head off large and unexpected expenditures once you're in the home.

After moving in, have a professional inspect your well once a year for peace of mind and to head off potential problems. If you choose to do this yourself, you can go to the EPA website: http://water.epa.gov/drink/info/well/index.cfm to read about the items you should be looking for. Perform a periodic water quality test by going down to your local health department and obtaining a water test kit. These are about $15 for a bacteria check and $35 for a full panel analysis. This will tell you if contaminants are getting into your drinking water.

Contaminant awareness. Your well should be at least 100 feet from anything that could introduce dangerous chemicals into your water. This includes septic systems, areas where you might use pesticides or fertilizers, and areas where machinery is kept that might drip oil or gas on to the ground. Keep animal pens and feed lots downstream and away from the well.

Environmental awareness. Try to use a minimum amount of fertilizer and pesticide on your property to reduce the chances that these chemicals will find their way into the groundwater around your well.

The well equipment. Check for a secure fit on your cover, and that the insulation is in good condition. Check the well casing for cracks and leaks. If you are not sure about anything you see, call a professional.

Shared wells. Many subdivisions have a shared well water supply. Make sure the community board or organization has a maintenance and service policy for the well and ask to see the results of the testing every year.

Clean water is a critical component of our health and vitality, but it is often taken for granted. If you own a private well, these tips should help you stay safe.

What You Should Know About Your Well

In network marketing they say, "Dig your well (establish contacts) before you need the water (need a job, favor, etc.)". Here in the mountains, though, we should say, "Take care of your well before your tap goes dry."

Amazingly, 85% of America gets drinking water from a public supply. That's why many of us, arriving in the mountains, take water for granted. Well? What well? Several Progress readers asked me this week about taking care of their wells. Since 80% of us living or vacationing here get our water from private and community wells, I thought I'd offer some advice.

Public water systems fall under the Safe Drinking Act. Regulated by the EPA (Environmental Protection Agency), public water supplies are subject to rigorous testing. Private systems are not regulated at all, except insofar as the well installers have to be certified. This fact may come as a surprise to you. It means that your well water can contain a variety of nasty chemicals and radioactive elements as well as bacteria, microorganisms, and parasites. If you get your water from a private or community well, here's what to do.

- Test the water every year. Visit your local health department and buy a test kit. A bacteria check is about $15, and a full test panel is about $35. If you have not done this in years, go for the full panel and then if all is well drop back to the $15 test each year. Follow the directions carefully to avoid a false positive. You can also order test kits online. If the tests come back showing deficiencies, call a local well company for professional help. The well servicer is trained to look for and find cracks, leaks, faulty valves, clogged filters, and review site problems such as drainage and contamination.

- Have the well and well equipment inspected every year by a professional. Sure, you don't want to spend the money – but remember - you're not paying a monthly water bill, right? Take less than half of that $500 or more annual cost and ensure your own health. Take the rest and put it in savings for the day when you need repairs.
- Stay environmentally aware. Don't over fertilize – where do you suppose the excess goes? The same is true for pesticides, fuels, and oils. Keep animal pens and feed lots downstream from your well.
- If the well is shared with others in the neighborhood, you should find out who takes care of inspections and maintenance. It may be one person or a group that rotates this responsibility. If you have a homeowner's association, they will be responsible for the integrity and safety of the well. Community wells should be inspected annually along with a full panel water quality test. If a community well fails or becomes contaminated, a lot of people will be affected.

So, the new saying is, "Care for your well and your well will care for you."

How to Keep Your Water Supply from Freezing at the Well

Most of us here is the mountains get our water from a well rather than a municipal supply. This introduces a set of maintenance chores that some folks don't know about, especially people moving here from homes that had a city water supply. The first time the pipes freeze at the well there is consternation and confusion. One Progress reader wrote, "We had just moved to a small cabin in Shooting Creek. I woke up on a very cold Saturday morning in January hearing 'Uh oh, no water,' from my wife. I had no idea what to do. Luckily there was a handyman next door and he placed a small heater in the pump house."

We have been lucky to not suffer the extremes that other locations around the planet have endured, from extraordinarily cold winters to flooding to raging fires. But that does not mean that we're not going to get some of these effects in the form of wild weather swings in the coming years.

Winter water supplies can freeze and cause problems throughout the plumbing system. Unless you turn off the heat inside the house without draining the pipes, the problems will be contained to the lines from your well to your house, the equipment in the pump house (or under your unrealistic looking well rock). Also unlikely but certainly possible is that the water line from the well was not buried deeply enough in the ground or soil was removed so that there is not enough depth for the line to stay unfrozen.

Typically, though, water freezes at the well head. This is the one area above ground that, if not insulated and protected, will freeze if we have many days in a row of very cold temperatures. Here's what to do if your water freezes at the well.

First, check to see if there is any insulation around the pump equipment if you have a pump house. If not, add an insulation box. Next, add a heat source. One way you can do this is to purchase two lamps. Install them near the pump equipment with 60 watt bulbs. Do not use CFL – CFL bulbs don't convert electricity to heat like incandescents do. You can install a thermostat switch that will turn on your lamps when the temperatures drop below 40 degrees. Or, when winter arrives, simply turn on the lamps. Why 2 lamps? In case one burns out.

Some folks use an electric space heater in their pump shed. I'm not enamored of this solution, because the thermostats tend to be unreliable, and it's a potential fire hazard.

A far more elegant and energy efficient solution, and one that works really well under those unrealistic looking rocks, is heat tape. This solution is available as a kit if you are feeling handy and want to do the job yourself, or you can simply call your well drilling company and ask them to "heat tape winterize my well."

The heat kit is composed of a length of electric heater cable with a power light at the plug and a built-in thermostat. Follow the directions carefully wrapping the pipes, and place the thermostat in the coldest area. Then insulate the pipes or purchase an insulation kit for your rock so that the heat is contained. The thermostat will turn the heat kit on when the temperature drops below 40 degrees. Voila! May you never hear the words, "Uh oh, no water!" again.

Should You Filter Your Water?

Should you buy and install water filters in your home? Whether you get your water from a well or a municipal supply, you may want to consider a water filter. While our water here in the mountains is generally clean and wonderful, contamination, especially in well water systems, can create taste and odor problems and the growth of bacteria and parasites. Contaminants can include heavy metals such as lead, copper, mercury, and arsenic; volatile chemicals (oil and gas); and pesticides.

The reasons for well water contamination include pesticide and fertilizer application close to the well field, keeping livestock near the well, a landscaping grade that leads contaminated water down a hill to the well, and where the ground water is contaminated by a septic system that is too close to the well or is leaking.

While the foregoing sounds dramatic, situations such as these are actually less common than you might think. Between code regulations and the professionalism of well installers and septic installers, there is usually plenty of safety built in to the construction and location of your well. Last week we talked about the importance of getting your well water tested regularly. If the test indicates a contamination problem, you should call a qualified well servicer to get this corrected. Many systems include a filter at the pump location. Your very first maintenance task is to make sure this filter is changed on schedule.

Do you need a filter inside the house? If your test results come back just fine, there is probably no need to add additional filtration. But some folks install filters anyway. Reasons include peace of mind, addressing a taste or odor problem, stopping occasional sediment, trapping rust from in house pipes or plumbing fixtures, and reducing the chlorine and fluoride in municipal water.

If you decide to install a water filter in your home you have a myriad of choices. Systems can be as simple as a counter top pitcher with a special filter to a whole house system that treats all of the water coming in to the house. The more complicated the system, the more maintenance it will require.

If drinking water and water for cooking are your primary concerns, then there are technologies that will provide filtration at your kitchen sink. A pair of in line canisters to treat water for odor and taste are straightforward to install (if you are not a plumbing wrench monkey I'd advise you to call a professional though) and will also remove chemicals and bacteria.

The next step up in under counter water purification is a 3-filter kit called "RO" or Reverse Osmosis. This technology produces very clean water that is free of bacteria, viruses, chemicals, sediment, chlorine, and radioactive substances. The kits can be installed by do it yourselfers or a professional. Maintenance is as simple as removing and installing new cartridges.

Since there are so many water filter products on the market, I recommend you do some internet research first if you decide to purchase a filter. A great resource is Consumer Reports. They do not accept advertising and are thorough in their testing. Their water filter buying guide can be found here: http://www.consumerreports.org/cro/water-filters/buying-guide.htm

Enjoy your "well" water.

How to Choose a Water Filter

Have you thought about the quality of the water you use in your home? Unless the water coming out of the tap looks cloudy or rusty, or doesn't smell right, we usually take it for granted. A large proportion of our population use a private or community well for our water supply. Some of the worst contaminants are tasteless, colorless, and odorless. They include the PCB (plastics) family, bacteria, lead, radon, arsenic, and a variety of sediments. In last week's column we talked about ways to take care of our well, and that you should get an inspection yearly by a qualified professional. Now we'll talk about how to choose a water filter to add to your system.

Start your search by deciding what you want your filter(s) to do. Types range from whole house units that filter out odors and sediment, to spin on charcoal filter units right at your faucet, to under counter units that process the water and deliver it to a special tap at the sink. The one to choose will depend on what you want to do. Do you want clean water for the whole house? Do you want water free of chemicals at your kitchen sink for drinking and cooking? Do you want to just filter out certain chemicals? After your well water test, you'll know what you need.

Whole House Filtration. These units consist of 2 large cartridges in series near the water source entry – often the basement. These systems should be installed by a professional. Typically, you will have a sediment filter first in line, then a taste and odor filter. Filters in these units should be changed every 3-6 months. The downside here is that if you do not change the filters regularly, they could actually introduce contaminants. They are expensive to install and maintain but will give you clean water for the entire home. These filters do not have the ability to remove all chemicals; you need a secondary filter for this if your tests indicate you need finer filtration.

Reverse Osmosis Filters. "R.O." filters, as they are called in the industry, are the best filters you can get for drinking and cooking. R.O. units install under your kitchen cabinet (or anywhere else you want one) and need little maintenance (filter once every 1-2 years, cartridges every 2-4 years). R.O. units have a downside though; they use water to make water; water goes down the drain for every cup that goes into the holding tank. This makes them impractical for uses other than drinking and cooking. R.O. units should be professionally installed.

Counter canisters: These filters are simple and easy to use. Check for the specification on the box for what it will and will not do. Make sure to follow the manufacturer's directions for cleaning.

Tap filter. These units screw in to the faucet at your sink. They are simple and handy, but like the countertop unit, check the specifications to make sure it will do what you need, and clean and replace filters on schedule. A dirty filter will be worse than no filter at all.

My advice is to research water filter brands by going online to Consumer Reports magazine (www.consumerreports.org). They do not accept advertising, and are thorough in their testing. They include specific ratings as well as prices. Another good source of information is the National Safety Foundation – www.nsf.org – for advice on certified water filters.

Tips Changing Your Whole House Water Filters

Many of us have wells for our water supply. In some cases the water needs no filtration, and in some cases the water needs an expensive treatment system. For those of us in between, two inline whole house filters (one for taste (carbon) and one for sediment (particulates)) work well. Changing the filters regularly is critical to the water quality we get at the tap.

There's a right way and a wrong way to do this. If you go to the internet and type "how to change Big Blue water filters," you will see several YouTube videos with someone saying, "Ok, this is really simple. Watch me." First, he forgets to turn off the water supply, then the filter housings are so tight he can't get them off, and then he spills all of the water on the floor because he forgot that they are heavy as they come off the housing threads.

Don't watch these videos unless you need a laugh. I'm going to tell you how to change filters without cursing or getting water all over the floor.

First, assemble what you need. The filter(s), a large bucket, a large trash bag, towels, paper towels, bleach, and plastic gloves. Find a hollow pipe ahead of time for leverage on the filter in case the last person over tightened it. Run a little water into the bucket to use for cleaning the inside of the filter housing(s).

Next, turn off the water coming in to the house. Then turn off the line right after the filters. Locate the red depressurization button on the top of the housing. Don't push yet! Place the trash bag around the filter housing and up over the top. Now push the button. The water that sprays out will go into the plastic bag.

Next, use the plastic wrench that came with your housing and turn the handle to the LEFT (Righty-Tighty Lefty-Loosey). Trouble turning? The person who last put the housing back on used too much force. You'll need to use the pipe to encourage the housing. Once you have it turning, stop.

Place the large heavy trash bag completely around the housing and up to the top where you should loosely fasten it so that it does not fall off. Through the bag, rotate the filter with one hand while you keep the other hand underneath the housing. When the filter comes off, it is going to tilt and spill water in to the bag.

Set the bag and housing on the floor and remove the spent filter. Pour an ounce of bleach into the housing and swirl around. Use a paper towel to clean and dry the inside. Use your plastic gloves to protect your hands. Remove the black sealing ring at the top and clean the groove. Use food grade silicone grease to coat the seal and reinstall. You don't have to change the seal for new if it looks fine.

Now drop the new filter cartridge into the housing. Tighten into the assembly, taking care to not bump the seal off. Go slow, and when it feels tight, use the wrench to go a quarter turn or just past hand tight.

Repeat if you have 2 or more filters.

Turn on the well side of the water and watch the housings carefully. If you have a seal rearrange itself, water will pour out the top of the housing. Hence, the bucket!

Turn on the house side valve. Go to a faucet in the home and open the cold side and let it run until you do not see any black carbon or particulates. Return to your filters and make sure there are no drips or leaks. Make a note of the install date.

You're done and there isn't even any water on the floor! Time for a YouTube video.

How Hot is Your Hot Water?

How hot is your hot water? When I inspect a home, I always test the hot water at the tap farthest away from the water heater with a calibrated thermometer. Why? Because I want to know 2 things: is the hot water heater actually working, and is the water heater producing water hot enough to deter bacteria, or "Legionella pneumophila", an organism that thrives in moist environments, such as water heater tanks. This was identified in 1976 after 34 veterans died during the American Legion Convention in the Bellevue-Stratford Hotel in Philadelphia. It then got the nickname "Legionnaires Disease."

Things just aren't simple anymore. Remember when you would simply turn on the hot water tap, wait for the water to get hot, adjust the tap, and jump in? The energy articles we read tell us to turn the hot water heater thermostat down to save money and now this article is going to tell you to turn it up to kill bacteria. What to do?

Get a thermometer and measure the water temperature at the tap farthest away from your water heater. If it reads 120 – 130 degrees, stop reading and go back to your hot shower. If it reads more than 135 degrees, you should probably turn it down to save money. If it reads less than 120 degrees, you should consider turning it up to prevent bacteria growth in the tank. If you are handy, you can adjust this yourself by going to the hot water tank and looking for the thermostat dial. It could be located on the outside of the tank, or it could be inside a cover near the bottom of the tank. If you don't want to tackle this, get your local handyperson to reset it for you. If you have a tankless water heater, bacteria is not a problem and you can set it wherever you want it. Dishwashers and laundry washers can benefit from higher water temperatures however; follow the recommendations for your appliance. Note also that water that is 140 degrees or more presents a serious safety hazard from scalding in addition to using more electricity or gas.

Another issue I run across on inspections is "that sulfur smell." This is an odor that emanates from the water as it comes out of the faucet. This smell is caused by bacteria in decaying plants, rocks, or soil and is picked up as water percolates through the ground. The byproduct of this decay is hydrogen sulfide. These bacteria do not cause disease, but who wants this smell coming from the tap? In this case, you should consult a plumber or go to your favorite home improvement store and determine what sort of filter set up is appropriate to remove the smell.

Hydrogen sulfide can also be produced inside your water heater by the magnesium rod placed in the tank by the manufacturer to prevent internal corrosion. The magnesium rod can be replaced with an aluminum one or removed completely to solve this, but you should consult a plumber, since removing this rod in the tank can void your warranty.

That's probably more than you wanted to know about hot water, but at least now you can answer the trivia questions about hot water settings, bacteria growth, and remedies.

How To Choose Your Next Water Heater

Buying appliances has become increasingly confusing. Instead of the relatively simple choices we had in the past, now we have to research what we need in order save money and end up with the right thing for our home and our lifestyle. Buying a new water heater, as mundane as it sounds, is now complicated by an array of choices. And we usually invest zero time picking a water heater because the purchase is driven by a family member's cry, "No hot water!" standing in a cold shower. It doesn't seem to matter what is on the plumber's truck: we want it, and we want it now.

If you think ahead, you can save money on the purchase and end up with a unit that will last longer. Here's what you need to know when you buy your next water heater.

There are tankless heaters, storage tank heaters, hybrid heaters, and solar heaters. In each of these categories, there are a range of types and sizes with prices from $300 for a basic electric storage tank unit to $3,000 for a gas condensing direct vent tank heater. To simplify our discussion, I'm going to compare units that most of us already have in our mountain homes or can install in our homes with the energy source we've already got: electricity, propane, or solar.

85% of us have an electric storage type water heater in our basement or garage. Given that our electric rates are relatively low (less than 10 cents per kilowatt-hour compared to a national average of 12 cents or more), electric tank replacements are a good deal. Things to keep in mind:

- If you're in the minority with propane, I would stick with this type but make sure the venting system is in good order.
- If you are replacing an electric tank, spring for the mid to higher grade units with a longer warranty. These units have larger heating elements, thicker insulation, and stronger corrosion resistance. You'll spend a little more but gain higher efficiency and longer unit life.
- If you really want another technology – such as tankless or solar, do your research. Be prepared to spend the money up front, and remember that tankless units will almost always use propane because they use a lot of energy in a short amount of time. Solar water heating can be great if you've got plenty of exposure where your house is located. Some of these technologies also offer substantial tax incentives.
- The more advanced the technology, the longer it will take to recoup your investment. An electric storage tank replacement will be the cheapest up front but cost a little more to operate each year, and the more advanced technologies, such as solar or tankless, will cost more to buy and install, but will cost less per month to operate.

If you have an electric water heater now and have to replace it, my personal advice is to get a hybrid unit with heat pump technology. Rebate checks of between $300-$400 are offered on these units, which are expensive ($1,000) but offer long life and excellent efficiency. Call your local electric company or visit www.energy.gov.

We hope it will be a very long time before we hear "No hot water!" again.

How to Make Your Water Heater Last Longer

Most of our appliances sit quietly and unobtrusively in our homes, delivering service reliably day after day and year after year. It is with great surprise that we get in the shower and realize that there is no hot water. "Why is this cold?" we say to ourselves with consternation. It is uncomfortable enough in the summer, but in the winter, it is tortuous to discover that your water heater has failed.

How long do water heaters last? Ten years is a general average, but it depends upon the quality of heater construction, and if it gets any maintenance. At the retailer, you will see a variety of warranties on the heaters. For example, a heater with an advertised "5-year warranty" will usually last 4-6 years, a "7-year warranty" about 6-8 years, and a "10-year warranty" 9 – 14 years. The advertised service life presupposes that you are taking care of the heater year to year.

What? Take care of the heater? You mean you have to do something every year with the water heater? Well, actually you don't. The only problem is that the service life will be less than advertised and you'll have to replace the heater sooner.

Why not perform some simple maintenance tasks on your water heater and get 30% more life from it? Think of it this way; instead of waking up to no hot water every 8 years, you can go 12 or 14 years without that cold water shock.

Of course, we've all heard the stories of a homeowner who does nothing for any of their appliances and the heater or refrigerator or washing machine has been functioning for 20 years. But don't count on it.

Here's what to do with your water heater to make it last. If you're not comfortable performing these chores, find someone who is, or you can call a service tech.

Find and read the manual that came with your water heater. I know this is basic, but there is a one page section that talks about care and maintenance. Tape this booklet to the side of your unit with the sales slip and warranty.

- Check the temperature. Keeping the heater at 120 – 125 degrees will minimize bacterial growth. I like 125. It's good and hot, and I know there is nothing growing in the tank.
- Check the temperature-pressure relief valve on the side of the unit by placing a plan under the PVC pipe coming off the valve and opening it up a few times to let several quarts of water out. Then watch to make sure there are no leaks from the valve. You are checking to see that the valve works. If it leaks, call for service.
- Drain the tank out every year. This will require a hose if you don't have a drain handy, and the instructions for the heater. Why? As water is pumped into your tank, dirt, sediment, and minerals settle on the bottom. As this sediment builds up year after year it will cause your heater to work harder and fail earlier.
- The last item – checking/replacing the sacrificial anode inside the tank every 3 or 4 years – is best left to a pro, but can double the life of your heater. Use the instructions or you can go online for detailed steps.

Here's to hot showers for many, many years without surprises.

More on Conserving Water

The lack of rain has been unnerving. The benefits – not having to mow because the grass has died – pales beside the concern that the well will run dry.

When I was growing up I remember my mom saying, "Turn that faucet off while you are brushing your teeth!" She was conserving everything she could – long before it became fashionable. It was just the right thing to do. City folk sometimes comment, "Well, you have a well, what's the big deal? You're not paying for it." True, most of us are not paying for our water, but it is still the right thing to do.

One person uses an average of 100 gallons of water a day, and the average family uses 400 gallons a day. That's a lot of water. Here are some reminders on how we can save water as we go about our day. I know you've heard many of these before, but they bear repeating.

Appliances. If you have a clothes washer or dishwasher that is older, and not labeled "high efficiency," then you are using 40 to 60 gallons of water to wash one load of clothes. High efficiency washers will use 10 to 25 gallons per load, more than a 50% improvement. An older dishwasher will use 15 gallons of water per load, versus 6 gallons in newer washers. And by the way, not using a dishwasher and leaving the faucet on to wash dishes means using an average of 20 gallons versus filling the sink – 10 gallons.

Toilets. Toilets are the biggest user of water in our homes. Most of us have installed the modern 1.6 gallons per flush units, but if for some reason you have really old toilets in your home that are still working, then you are flushing away 5-7 gallons of water every time you use it. When it's time to replace your toilet, you can now get a unit that uses only 1.28 gallons per flush, and yes, it actually works.

Faucets and Showerheads. It goes without saying that you should pay close attention to when you leave the water running. These devices deliver between 2.5 and 4 gallons a minute. Easy ways to save include turning the faucet or showerhead off when you're soaping up, and/or replacing them with low volume devices. I hear what you're saying: "I hate low volume showerheads, I want WATER! The more the better." I understand this, but you can now purchase low cost ($20) 1.85 GPM fixtures, including the one I bought last year – Delta adjustable flow with "H2O Kinetic Technology" which makes you think you are under a Hawaiian waterfall. Well, not quite, but much better than the old models.

Fix leaks. A toilet valve that leaks can waste 300 gallons of water a day. A slow faucet drip is 3000 gallons a year. The average household's leaks can account for more than 10,000 gallons of water wasted every year, or the amount of water needed to wash 270 loads of laundry.

Other ideas for saving water include taking shorter showers, not rinsing dishes before putting them in the dishwasher, waiting for full loads before running washers, and getting a rain barrel for the garden. All it takes is developing awareness that water is one of our important resources. We'll be all set now if it would proceed to fall in abundance from the heavens.

Three Common Roof Leaks

When was the last time you took a good look at your roof? Quick, tell me what your roof is made of. Is it metal? Shingle? Shake? Asphalt? Ok, some of you don't know. That's OK. But it tells us something – it tells us we take our roof for granted. We know that a roof covering should last a long time, and, like our car, it should keep going without any problems. We know that's not entirely true with our car, and it's also not entirely true with our roof.

If your roofer told you he installed a "20-year roof" does that mean you can forget about it for 20 years? Sorry – but that is not what it means. It means that the material itself may last for 20 years – but the rest of the components, such as flashings, boots, soffits, and wood trim – will need attention long before then.

Most of us have either a composition shingle roof (asphalt or fiberglass) or a metal roof here in the mountains. Both types are very durable and if installed correctly will last the life of the warranty. What would cause early water leaks into your home?

Most early leaks come from three things. These are easy to check, and if you don't check them every year, you may find yourself with a water leak in as little as 5 years after installing a new roof or buying a new home. I always look at these three things when I inspect a home. Last week I found one of these leaks in a home that was only 3 years old.

The first and most insidious leak comes from the plumbing stack vent boots on your roof. These are the PVC (white) pipes you see coming out of your roof. They allow air to exit your plumbing system. These pipes run inside walls, and when the gasket around the pipe outside (where it meets the roof) deteriorates, guess where the rain goes? Yes – into your wall. It may take some time for you to discover this nasty leak source, and when you find it, it may not be apparent where the water is coming from. Take a look at these stacks every 6 months to a year – a pair of binoculars is usually all you need – although I have seen them look fine on the sight line from the ground and cracked on the roof upslope. If you see deterioration, have a qualified roofer replace the gasket or flashing.

The second most common leak is around the chimney flashing on the roof. The chimney is a large structure exiting your roof and you know that water will find an entry wherever it can. These leaks are more obvious – you'll usually see the leak at the interior ceiling joint where the stone work meets it or meets the walls. If you see this, it's time to find a good roofer to re-seal or replace the flashing around your chimney. Make sure you don't just send Hank or Hanna the Handyperson up there to do this unless you're sure he or she knows what they are doing. Using a lot of silicone caulking on areas that should be flashed with metal or rubber will usually stop the leak this week but will make the problem worse over time.

The third most common water entry into your home is from clogged and overflowing gutters at the roofline. When these fill up due to a clog or backup in the downspouts, water splashes back on to the trim where the shingles end. The water gets under the shingles and rots the fascia, and can even cause water to run inside the walls. If left like this for a long time the moisture will actually rot out not only the fascia but also the soffits at the overhang – now you have a big repair bill.

So, the next time you go outside, take a GOOD look at your roof and cross that one off your spring checklist.

How Much Do You Know About Your Septic System?

Life in the mountains for most of us means a home with a septic system. Like our other appliances, septic systems are extremely reliable. However, they can, and do, fail. Failure happens when we totally forget about the system, thinking it is a permanent, working appliance for the life of the home, or we do not understand enough about the system to steer clear of trouble.

When I was a home inspector, I'd hear a variety of the following after telling the buyers that the home had a septic system:

"What exactly IS a septic system? It doesn't need attention, right?"

"If it fit's down the drain its ok, right?"

"It's a great way to get rid of kitty litter."

No, no, and no!

Septic systems maintain a delicate chemical balance that allow bacteria to clean the wastewater and carry it to a drain field where more bacteria cleanse the remaining water. The inorganic materials accumulate in the tank, and this is why the tank needs to be pumped out periodically.

Here's how to save money and prevent trouble.

If it's not organic, it should not be put in the septic system. Minimize your use of disinfectants and bleach. One product I like is Lemon Soft Scrub for the bathroom. NEVER put the following items down your drains: paint, oil, drain cleaner, solvent, pesticides, cat litter, coffee grounds, swabs, cigarettes, sanitary products, and grease or cooking oil. If you have a disposal, use it only to get rid of plate debris after placing large items in the trash.

Conserve water. Running lots of water through the system in a short period of time reduces the pre-treatment phase in the tank, so the system is not as efficient as it should be. Spread laundry loads out; don't let the faucet run if you're not using it; repair any water leaks in your home.

Don't drive a vehicle over your septic tank area, and keep grass on the drain field. Don't plant in the drain field, and don't drive over the drain field. Don't know where the system is? See your building or health department; they will usually have the permit information.

Have a qualified contractor pump your system out every 3 years. If your system receives minimal use (you're not here all the time or there are just a few people in the home) you may be able to go a lot longer than this. At the least, get the system inspected every 3 years.

Early signs of septic-field problems include patches of bright green grass; turf that feels spongy when you walk on it; toilets, showers, and sinks that take too long to drain; sewage and/or smells near the leach field; and sewage odors after a rainfall. If you notice any of these signs, have your septic system inspected immediately.

Finally, don't bother using septic system enzyme or bacteria additives. Engineers and sanitation professionals say that these additives can actually do more harm than good.

With periodic pumping and attention to what goes down the drain, a septic system can be trouble free for 30 years or more.

23

Upgrades, Remodeling, and Warranties

Should You Tackle That Fixer-Upper?

Homes that need care and repair are in plentiful supply since the last recession. These troubled homes can be a great bargain. However, most real estate professionals steer clear of these damaged homes because the unknowns carry high risk. Should you consider buying one of these fixer-uppers? Here's how to decide.

Can you deal with the unknown? If you are comfortable knowing that there can be hidden damage in your fixer-upper and that it is impossible to calculate how much things will cost going in, then you might not regret the decision.

Are you construction savvy and handy with repairs? If so, it will make things easier and you'll be able to determine what contractors to hire.

Can you handle stress well? If you have a balanced attitude knowing that your fixer-upper will produce some surprises and take longer (and cost more) to complete than you think, you may survive the experience.

If you decide to buy a damaged home, follow these tips to reduce the problems you'll have to deal with.

Spend time finding out what is wrong before you buy. Just getting a home inspection will not do in this case; you will need to take the home inspection report and follow each "clue" or issue using thorough investigation and experts to determine the extent of the problems. For example, if the inspector flags an area of moisture or mold on a wall, you should have a contractor perform thermal imaging using an infrared moisture detector. This will improve your chances of finding everything that is wrong and being able to budget accordingly.

After listing the issues, determine what you can do and what you will need a pro to do. Code will dictate finding a licensed contractor for systems installation and repair such as HVAC, electrical, and plumbing. You can work on almost everything else. Consider getting professionals for the items you're not comfortable with – they will save you money because they will spend less time on the job versus you stopping and starting every time you realize you are missing something.

Become a great negotiator. Repair costs on a fixer upper take on a life of their own if you don't push for price breaks. Contractors know you are trying to save money and most will give you the best deal they can, but you should still ask without being obnoxious.

Over budget for repairs. This way you might be pleasantly surprised.

When I was an inspector I helped dozens of people consider problem homes. I was amazed at what folks were willing to purchase knowing how many problems existed. One of the worst ones was a home with a "creek" running through the crawlspace and missing supporting structure at the foundation causing the floors to slope, the windows to crack, and the roof to leak. When I returned to the area 2 years later, the home's foundation and structure had been completely re-built and the owner said it is her dream home now.

Best Home and Yard Remodeling Ideas

As we approach the delight of summer months in the mountains I am sure you are making your outdoor to-do list. Many of the items are repairs that have to be done; but I hope you also included some wish list items. If you didn't, you should. Why not take these gloriously beautiful days to build something new?

Real estate professionals will tell you that a large portion of every dollar you put into renovations and improvements will come back to you when you sell your home. But the best reason for thinking about improvements is you. The updates, changes, and rearrangements you make will be fun for you right now.

If you want to combine both reasons – personal enjoyment and resale value – then you will have the best of both worlds. Here are the top home remodeling and improvement projects according to realtors and contractors.

Update your home's entry. This is at the top of list for realtors, who remind us that the first thing potential buyers see is the entry to your home. Walk out your front door and into the street or yard and spend five minutes looking at it from every angle. What do you see? It should be clean, inviting, and reflect care and attention. If you usually drive into a garage you will not have the same perspective as guests. Make a list of the improvements. This could be adding or refreshing the path to the front door, installing a new front door, painting or staining, and cutting back plant growth. Real creativity comes in if you decide to change or add landscaping elements such as a recirculating waterfall, dry creek bed, flagstone path, plants, and solar lighting.

Deck and porch additions are always a high pay-back item. U.S. News and World Report magazine reports that this one upgrade to your home can return over 95% of what you spend now in a future sale. It does not have to be fancy composite decking – pressure treated pine will do fine. Take note however, that spending a little extra to have your contractor use screws instead of nails will improve longevity and reduce maintenance.

Upgrade kitchen materials and/or appliances. We spend a lot of time in the kitchen. Take a look around yours and make a wish list. If your appliances – oven, refrigerator, dishwasher – are more than 10 or 12 years old, upgrading can bring higher energy efficiency. Granite countertops are at the top of buyers' wish lists, but modern laminates deliver a similar look and feel for 15% of granite's price. New cabinets will pay for themselves on resale and it's fun to add a new look and extra functionality to a kitchen.

If your home's entry includes a long driveway, take a good look at the entrance. Some refreshed landscaping will make a big difference in appearance and appeal.

Other projects yielding good resale value as well as current enjoyment include adding a garage, adding a sunroom, adding a bathroom, adding a home office space or den, and adding a cohesive landscape design with pathways, plantings, and lighting.

So, don't despair thinking all your time this summer will be spent mowing and fixing – think up some fun projects that everyone will enjoy for years to come.

Choosing a Contractor: Avoid These Traps

Sluggish home sales combined with reduced earning power in today's economy mean we have to do more with less. Each year you may say to yourself, "Things are getting better; perhaps next year we will buy a bigger home or build a smaller one with all the things we have been thinking about." These options may be impractical or out of reach. If this is the case, you may simply need an upgrade or room makeover to create house harmony.

Unless you are going to do the work yourself or already know a contractor, you will be on the lookout for someone to perform these upgrades. How do you choose a home contractor? As you embark on this important search, consider avoiding the following traps.

Trap #1: Don't worry about licensing or insurance. We're in the mountains and everyone is friendly and honest, we know that. Your neighbor is talented and competent, we know that. But what if you end up in that 5% group where something goes very wrong and now what you thought you wanted and said is a "misunderstanding" and your "contractor" leaves you with a botched job? My advice: make sure your contractor is licensed and insured.

Trap #2: We don't need a written statement of work; a handshake will be fine. Similar to Trap #1, 90% of the time everything will turn out as you expected, but if you fall into the 10% where the "misunderstanding" rears its ugly head, you will have to accept what you get. My advice: ask for a written scope of work with the details of work delivery, timing, and cost. Professional contractors consider this to be not only routine, but highly protective on their end as well. If you encounter a contractor who thinks an agreement is extra, I'd be concerned.

Trap #3: Pick the first guy or gal you come to in the phone book. Phone book? Do we still have phone books? My advice: wherever you look – internet, newspaper, yellow pages – do your research. Why not get the best work possible? Although it's more time consuming, I also suggest that you obtain referrals from prospective contractors. Although some of these referrals may not be impartial, they will usually be honest about the contractor's strengths and weaknesses. You can also check Angie's List, a vetted review of contractors online.

Trap #4: Expect the best without communication. Once you have the scope of work and you've told the contractor what you want, that should be it, right? Wrong. Advice: you cannot communicate enough. You should be checking in on things often, and providing frequent feedback. Even with a scope of work, you could get to the end of the project and suddenly realize that it's not what you wanted.

Trap #5: You don't have to like your contractor to get a good job. Why should personality have anything to do with accomplishing upgrades in your home? If they're good, they're good; you are not paying them for their personality. Advice: Your contractor can be licensed, insured, have 25 years of experience, but if you don't get along with this person, the job is not going to go well. This rubs both ways. If you don't respect and like your contractor, they are not likely to respect and like you either. Communication, and workmanship, will suffer.

Happy remodeling.

Quick Tips for Home Remodeling

With new home construction prices still out of reach for many homeowners, an addition or remodel to an existing home can be a great investment and provide the excitement of something new and fresh. Remodeling a portion of your home can also pay for itself in a future sale. High return areas include kitchen, baths, porches and decks, and landscaping.

However, home remodeling and additions are stressful. From strangers in your living space to money worries, it can present a difficult situation. Throw in schedule and cost over-runs, and you may wish you had not started it at all. The good news is that 80% of the people who remodeled say they are glad they suffered through it.

Ready to begin? Here are seven quick tips to navigate the process smoothly.

- Get everyone in on the advice. Hold a brainstorming session with your family or friends to come up with the best ideas. Write everything down and then consider the pros and cons of each idea. This also works for designs. From bringing the laundry room upstairs to building a bump-out dining area, this session is fun.
- Choose a professional contractor. If this is your line of work, then have at it. Otherwise, I recommend you use licensed professionals. You may pay a little more for this level, but you'll all but eliminate the nightmares of shoddy and unsafe work, an injury at your home that is not covered by insurance,

financial irresponsibility, and not providing follow-up service and guarantees.

- Consider simple changes. While you might wish for a large addition, it's expensive. Consider removing a non-structural interior wall and unnecessary doors to open up space for a great room, a much less expensive remodel. Use the guest room as a double duty media room or game room. Expand the deck or porch and enclose half of it for year-round living.
- Add 20% to your budget. Whatever the remodeling plan is, help head off financial surprises by upping your budget allowance. Maybe you'll have something left over for decorating.
- Don't skip the contract. A handshake deal is ok for simple chores when you know who you are dealing with, but having a detailed written agreement for large jobs is critical. Read the contract carefully, ask questions, and make sure everyone has signed it. This is a legally binding agreement if things go wrong.
- Don't fall for the Fad of the Day style or color in your decorating. Stick close to the colors and styles you have already, but move a few creative steps out. Completely changing the theme of one area will make it appear disjointed from the rest of the home.
- Get rid of junk. A lot of "stuff" around the house makes the spaces look small. Why not clear things out just before the holidays? For the things that you cannot part with, build bookcases on several walls with doors on the lower shelves. You'll be amazed at how much hidden storage you will get, and just how much space is available on the shelves. Shallow shelving works wonders in a small kitchen too.

We all enjoy change from time to time. If you just need a fresh perspective and don't have much cash, consider fresh paint and rearranging the furniture!

Get the Most Bang for the Buck on Your Home Upgrades

As any real estate professional can tell you, buyers can be fickle about what they want in their next home. Whether you are considering upgrading or renovating areas of your home for yourself, or to make your home more appealing to a buyer, follow the suggestions below to get the biggest bang for your buck. According to Consumer Reports, each of the ideas below may raise your resale value by 3-5 percent.

- Hardwood floors, up to date heating and cooling system, and recent re-roof. All generations of buyers are taking a hard look at the time it takes to maintain a home, and interestingly, the younger buyers do not want to spend a lot of time maintaining their home.
- Increase the amount of living space. If you've got a home that was never completely finished – from a side room to an unfinished basement, spending the money to finish these areas will pay off for you both now and at resale time. Keep in mind that younger buyers also say they want a dedicated laundry room on the main floor.
- The kitchen is still where the action is. The 80-million-member Millennial group, comprised of those people born approximately between 1980 and 1995, represent one third of the entire U.S. population and are specific in their wants. An open floor plan and a modern kitchen are at the top of their list. This means solid wood cabinets, stainless steel appliances, and granite or quartz countertops.
- Decks and patios. All generations love the relaxation and de-stressing that the great outdoors brings, especially here in the lovely landscape and wildlife of our special mountains. Easy to maintain areas that are simple will be the most appealing. Consider adding a deck if your home does not have one, or

a small patio with a fire pit in the back yard to increase your home's appeal and value. You will love it too.

- Wireless security, locks, cameras, and thermostats. The life of a Millennial is totally connected – and they love wireless gadgets. Wireless security and convenience – all connecting to a smartphone – are very inexpensive to add to your home. Cameras are $200 each; locks about the same; Honeywell connected thermostats are about $275 each. This is a small investment with a big appeal. You might even enjoy these newfangled gadgets yourself!
- Whole-house generator. Electricity means a lot to younger generations who have not really had to go without it. The appeal of a generator that can automatically switch over seamlessly and with little worry in maintenance, is a big plus to buyers. If you decide to make this upgrade, you may find yourself glad that you did it in the next storm. This upgrade will cost you between $4,000 and $12,000 installed, depending on the size of your home.

So the next time you think about upgrades, think carefully about what will not only please you and your family, but what will give you the biggest bang for the buck on resale.

Do You Want the Warranty with That?

The Wal-Mart checkout line seemed to stretch on for miles across the store, with yellow "Do Not Cross" barriers herding shoppers with carts piled high on Thanksgiving Day. This was the fourth line I had waited in, and I was getting antsy. As we slowly moved to the checkout counters, I kept hearing the cashier say, "Do you want the warranty with that?"

"Why should I get it?" asked the woman in front of me.

The cashier replied, "Um, well, if it breaks, you can get a new one."

As the customer stood there with a quizzical look, the cashier handed her a card that explained what the extended warranty would do for her if she bought it. Meanwhile, the line behind us fidgeted and waited patiently while the shopper consulted the card in hand and then her companion.

"Martha, what should I do? Should I buy the extra warranty?"

"I don't know! It sounds like a good idea."

"Ok, the shopper said, "I'll get it."

A sigh of relief issued from the line of customers waiting, and the line snaked forward once again. Meanwhile, under my breath, I said "No! Don't do it!"

Here are five reasons to NOT buy the extended warranty.

- You already have a warranty. All brand new purchases already carry a warranty. The question you should ask yourself is, "will this product fail inside the warranty period or after?" The odds are high that if the product stops working it will fail as soon as you plug it in or very shortly thereafter.
- Electronics and appliances are extraordinarily reliable. You would be better served to save the extended warranty money for the rare day when you need to pay to have something fixed.
- Extended warranties are expensive. What the sales people do not want you to know is that the extended warranty plans are pure profit for the store, and many plans pay the sales person a commission to push the plans. I'm not against businesses making a profit, or salespeople making commissions, but wouldn't you like to know these details before you buy an extended plan? Adding an extended warranty to a product can cost half what you paid for the item.
- Better deals after checkout. If you really do want an extended warranty, consider not buying it at checkout. Check online

with Square Trade or Protect Your Bubble for less expensive after purchase warranties.

- Your credit card may cover it. Many credit cards have a product protection plan built into their benefits that cover items for one year after your regular warranty expires.

There are times when you SHOULD consider an extended warranty. Peace of mind and less aggravation come from getting the 3 year we-will-come-to-you-immediately warranty on a desktop computer or a piece of exercise equipment.

What about the "Accidental Protection Plan?" Unless you regularly drop or run over your phone or laptop, these are expensive. Put the money in your savings account and over time you'll be ready for the catastrophe.

When I finally got up to the Wal-Mart cashier on Thanksgiving with my cordless dirt devil, I said "No extra warranty, thank you."

Picking the Right Ceiling Fan

In the heat of the summer it's nice to have a comfortable porch chair, a cool drink, and a ceiling fan. There's just something about this combination that shouts "relax!"

Ceiling fans don't change the temperature, but they do cool us by evaporating moisture on our skin, taking away that clammy feeling. Fans do a good job of keeping humans more comfortable when it's hot and reduce our need to lower the thermostat on the air conditioning. And ceiling fans just look cool, even when they're not running.

But choosing a ceiling fan can be confusing. The last time I looked, there were about 60 ceiling fan companies in the United States alone. Where to begin? Use these tips.

Style. Go to your favorite home store and look around. Take your time and look at everything. Wood? Metal? Color? Antique? Modern? Think about the décor in the room where it will go. Make sure you're happy with the combination of style and color. For high tech, look at the curvy bamboo blade fans made by Haiku Home. The only problem is that they are expensive.

Size. Use this quick formula: Measure the square footage of the room – length times width – and then divide this by 4. For example, a 10 by 15 room equals 150 square feet. Divided by 4, this equals 38 inches of blade span. Move up or down in span to account for ceiling height.

Blade material. Look for solid wood, metal, or plastic. Wood composites can take on moisture and warp over time. Putting up on a porch? Choose ABS plastic.

Motor. DC or AC motor? While Alternating Current motors are smoother, Direct Current motors are quieter and more efficient. DC motors are more expensive but save on electricity, while AC motors are less expensive. It likely does not matter if the fan is in your price range and you're not concerned with how much electricity it will use.

Efficiency. You may think that the more blades a fan has, the more efficient it is. Not true – what makes a fan efficient is how much air it can move at high speed, or "CFM" – Cubic Feet per Minute. High CFM fans will have robust, quiet motors that last a long time. To see efficiency ratings of fans, go to www.energystar.gov and type in ceiling fans.

Price. Once you know what you want, all that's left is to pay for it, right? If you've done your research you know that there's a very large range in the prices of ceiling fans. At this point my advice is to go to Consumer Reports (consumerreports.org) and look at the ceiling fan ratings. Another great spot is amazon.com, where you can read "real" reviews from purchasers.

Finally, you'll need to figure out how to install the fan. If you are a handyperson, and the fan is going into an existing light box, it's not a difficult job, but make sure you follow all of the directions, including the safety warnings. If it is an all-new installation, I highly recommend a professional for the job, since a secure mount must be installed and an electrical line will need to be run.

Consider buying a fan with a remote control so you don't have to mess with chains. Some fans allow you to control the fan with your smart phone. How "cool" is that?

What You Need to Know About Product Recalls

Soon we'll be gathered around the Christmas tree, looking at wrapped gifts. A short time later, we'll be looking at tinsel, paper, boxes, and toy parts strewn across the floor.

"What happened to the assembly instructions," you exclaim, as you try to determine the order of construction on the remote controlled Humvee. "They are here somewhere," a family member says, looking over the jumble of paper as the cat jumps out from under a pile of tissue paper. As you look through the mess, you encounter several warranty mail in postcards for the toys.

"More junk," you say.

Hold on. Let's talk about product recalls and safety. This is the last thing on your mind as you unpack new products throughout the year, not just Christmas time. Consider holding on to these postcards. While you may not care about registering the warranty, there is a reason for letting a manufacturer know that you bought something, and that reason is called a recall.

If a manufacturer discovers a defect, especially a safety problem, they try to let consumers know. Examples of recalls or safety alerts over the last year include 54,000 Ryobi 18 volt Lithium battery packs that can burst while in the charger; Belvita Breakfast Biscuits that could contain fragments of metal mesh, and 5,500 motor scooters sold at Walmart that can accelerate without warning. A recall just last week involved small polymer beads for children that expand when put in water. I don't need to tell you what these beads do if a child mistakes them for candy, but it is life threatening.

Recalls over the last five years have included bursting toilet flushing valves, disintegration of parts on leaf blowers, refrigerator doors falling off, and battery fires.

What can you do about this? Here is a quick set of tips to help you stay safe.

- Keep the registration cards that come with products. Throw these into a file or into the drawer with the bills, and fill them out when you're doing other paperwork (an address stamp makes the chore short). This lets the manufacturer know how to get hold of you, and also makes using the warranty much easier if you have a problem. You can also register most products online.
- Check the products you already own by going to www.recalls.gov. If you are having a problem with a product, you can report it on www.saferproducts.gov.
- Research what you are planning on buying. We assume that sellers are aware of recalls, but this is not always the case, especially when it is a resale. It is illegal to re-sell a recalled product, but resellers do not take the time to check for recalls, so it's Buyer Beware.

Sure, it is more work to send the postcards in and check on potential problems. But it's also an important extra layer of safety in an uncertain world.

More on Extended Warranties

Are you in the buying mood? As shoppers scurry, finding the best deals in the holiday season, bargains abound. It is a good time to find the items you want, and since it's that gift giving time of year, it seems easier to justify the purchases. However, don't spend more money than you need to! There's a hidden trap that you need to know about: extended warranty plans.

As an electronic technician in the 1980's, I sat at a test bench running products through a testing sequence to see if the electronic components would fail. After that test, I would then run the unit for 20 hours in ovens to see if they would fail over time. What I found out from this extensive testing was that if the piece of electronics did not fail immediately, it would not fail at all for a very long time.

What this means is that if you buy a piece of electronics – an LCD TV for example – and take it home and plug it in and find that it is working, the chances are about 95% that it will continue to work for many years. I am not saying it will never fail, and of course there are the exceptions, but if you stick with the statistics and averages, you will save a lot – I mean a lot – of money by not falling for the sales pitch to purchase an extended warranty for the TV.

Salespeople may not tell you that the store makes a big profit on these warranties, or that they get commissions from them. They may not tell you that the plans overlap the manufacturer's own warranty rather than getting tacked on when the manufacturer's plan runs out. They may not tell you that there is no guarantee the warranty provider will be there when you need them, or that they will provide competent service. They may not tell you that the credit card you use to buy the product may have an extended warranty provision attached to it for free.

If you buy 10 appliances or pieces of electronics over the coming year and you also purchase a service plan for each one, you may have spent an extra $1,000 or more on these warranties over the price of the products.

My advice is to never buy an extended warranty for any piece of electronics. Each year take half what you would have spent for the warranty and put it into your savings account. If you finally have a failure in a product outside the original manufacturer's warranty, you will have a pile of money for the repair or replacement.

In some cases, you may decide that an extended warranty plan along with the promise of in-home service is attractive enough to spend the money on a plan. Computers and large appliances are good examples. These plans are not cheap, but you may not want to drag your 3-year-old PC down the street to the computer store. But if you really want to save money, the law of averages says that you will not have a problem in the warranty period. Let's go shopping!

24

Secret Hiding Places and Other Fun Stuff

House Sounds That Require Attention

About 80% of the "ghost" sounds you hear in your home are harmless. These include cracks and pops from heating and cooling of materials that expand and contract at different rates, wind blowing through vents, creaking floors, and tapping sounds in your water pipes. Unless, of course, you actually have a ghost in your house.

Other than the ghost, there are five sounds, however, that should be investigated – they could represent serious problems that require attention. I'll talk about each one, and how much you should worry.

High pitched squeals. Unless you're keeping pigs in your basement, squealing is definitely not a normal sound in your home. When you track this squeal to its source, you are likely to find a belt or bearing in your heating and cooling system that has failed. Try turning the system off for 15 minutes and turning it back on. If the squeal returns, shut it off and call an HVAC professional. Check your filters to make sure they are not clogged, making things worse.

Scratching or scurrying noises. If you hear either of these noises in your walls or ceiling . . . it's either that your hamsters got out, or some other little animal has found a way into your home. This investigation is best left to a professional pest control person who can determine what the vermin is and get it out. The longer they stay . . . the more they enjoy your hospitality and the bigger the mess will be.

Humming or buzzing sounds coming from switches or fixtures. If you hear these from any electrical device except your blender or coffee bean grinder, investigate. This could mean that a wire is loose somewhere in the outlet, fixture, or panel. This is a fire hazard and should be checked out fast – call an electrician. If you're handy, check out outlets (turn power off!) but if your electrical panel is humming, don't touch it - call an electrician immediately.

If you hear humming coming from light fixtures, it usually means that the bulb is incompatible with the switch (usually dimmer switches), or in the case of fluorescent lights, that the ballast has gone south. While these noises are not emergencies, it does mean that the bulb's life may be shortened or that you will not get rated output. Make sure that the bulb is rated for dimming – it should say so on the box.

Water heater crackling or popping. If you have not drained your water heater tank in a while, sediment has built up in the bottom of the tank and is actually boiling when the elements come on. While this set of noises is not an emergency, the life of your water heater will be considerably shortened. If you hear this sound, draining and flushing the tank should cure it. Call a plumber.

Dripping noises coming from the fireplace. Is it raining and is water dripping in your fireplace? Not an immediate emergency, but over time this water, which is making its way through a failed flashing or cap, will rust your firebox, creating a fire hazard the next time you throw wood in and the floor of the box drops out.

Fortunately, the benign sounds in a home outnumber the bad ones. If you hear the bad ones – ghost or no ghost - take action!

Things That Go Bump in the Night

We've all heard noises in our homes that we've wondered about. Usually the time to hear them is right after getting into bed. As you lie there willing a restful sleep to envelope you, you hear something. What was that noise?

The first night in a newly constructed home, I was on a foam mattress in a sleeping bag on the carpeted floor of the bedroom. I was so excited to be in the home, I was there before the furniture was. I turned out the light, and as things quieted down I heard the sound of dripping water. "Uh oh," I said aloud to myself. "There must be a leak in the wall!" I turned the lights back on and looked around. "It must be trickling through the wall and into the basement," I said. Then the noise stopped. A minute later the drip-tap-drip sound started again. Perplexed, I went out into the hall and down the stairs into the basement. For 15 minutes I searched everywhere, and could not find a leak.

I went back upstairs, exhausted, deciding I would ask the plumber about the noise the next day and determine if it is a problem or not. I drifted off to sleep with water streaming through my dreamscapes.

There are a myriad of household noises that are perfectly normal, and a handful of noises that are trouble. Here are the three most common problem noises in plumbing systems and what to do about them.

The drip or tap noise in the walls. What I was hearing in our new home was the sound of CPVC (plastic) water pipes expanding and contracting on structure or against fittings as warm water moved through cold lines. Convection can cause water movement even when you don't have water running through a fixture. This sound is common and normal if the plumbing lines were not insulated or secured in special sound damping pipe hangers. However, if you are not sure, you should turn your water off at the main shutoff and then listen and explore to rule out an actual water leak.

A bang noise turning off the faucet. This can be quite loud, and is commonly known as "water hammer." The sound comes from water that is in the system coming to a quick halt as you turn off the fixture.

If you routinely hear this, you should investigate. Air chambers are typically built into a plumbing system to act as "shock absorbers" and these may be clogged. The other major cause of water hammer is pressure that is too high in the system. Over time, water hammer can cause failure of valves and fittings, and even cause the pipes themselves to burst.

Cracking and popping sounds coming from the water heater. When your water heater was installed, it came with a manual. In the manual, a maintenance section advises the owner to flush the tank several times a year to remove sediment. Have you ever done this? Of course you haven't. No one does this. But the loud crack and pop sounds come from boiling sediment in the bottom of the tank. The sediment acts as an insulator, forcing your tank to work harder, and fail earlier. If you hear this sound, draining and flushing the tank should cure it.

More Sounds: Is Your Home Haunted?

Most "haunted house" sounds are not dangerous and can be easily explained. But try telling this to someone who is alone in the house trying to fall asleep. In scary movies, these same sounds tell us that something bad is about to happen. The "haunted house" moniker is usually applied to homes that are old, large, and have metal piping, radiators, wood doors and floors, and no insulation. After 50-100 or more years of sitting in one place, the wood framing, fasteners, metal, glass, window frames, door hinges, and ductwork are all a little looser.

Here are the top six "haunting" sounds and what you can do about them besides holding a séance.

Creaking floors. All older homes have areas where the floorboards move. This is caused by fasteners loosening. What to do: You or a handyperson can install additional fasteners and shims. If you have a crawlspace this should be straightforward; if you have to perform this task from above, then consider buying a special kit from the hardware store with breakaway fasteners.

Doors opening and closing by themselves. Having doors open and close by themselves can be unnerving but is probably not a ghost coming in. The cause is usually a door frame that is loose with a door out of level, along with air movement. The fix for this is to repair and secure the frame, making sure that you have a level door. A very simple fix is a door stop.

Scratching sounds in the attic. This is a common sound at night if you have uninvited guests making their home in walls or attic spaces. These guests include mice, squirrels, birds, and raccoons. On one inspection as I popped my head up into the attic, a dark birdlike creature swooped past my face and I nearly fell off the ladder in to the room below. Bats! If you discover any of these animals in your home, you will need to get rid of them before you seal up their entrances . . . or you will have strange odors in addition to the strange sounds.

Moaning and clattering. Rather than spirits, these sounds are more likely to come from air blowing through vents and walls. Typically, you will hear these creepy sounds when it's windy outside. The cure is to methodically identify where air is entering your home, and to check the vents to see if one or more ducts have a hole or the flapper is missing or not closing.

Whistling and air noise. While the whistling sound can also be from air moving in around windows and doors, hearing this sound close to you probably means that a filter is clogged in one of your heating and air conditioning vents and the air has decided to go around the filter. Replace the filter and check the others and this noise should disappear.

Bumping, popping, cracking. These noises, especially prevalent in the morning as the sun is heating your walls and in the evening as the sun is retreating are the sounds of your home expanding and contracting – concrete, wood, piping, ductwork, and siding – all moving at different rates. This set of noises is normal.

Now that you understand common home noises, the only thing you will need to worry about is the creature under your bed.

How to Assemble Stuff

Over the years we've all been called into service in one way or another to assemble something, sometimes at the last minute. More than one Christmas Eve my sister has called me asking if I can run over and put the doll house together, or the patio chairs, or the computer, or the barbeque grill. All I can say is she is lucky to have a sister who loves assembling things.

When I ran a bicycle shop in the 1970s, one of the things I looked forward to was putting the bicycles together. Some of my customers did not want to spend the money for a really good bike ready to roll off the floor and wanted to get the budget deal at K-Mart in a box. I advertised that I'd assemble the bike in a box for $10. Christmas Eve I was always busy and open late so folks could come get their children's gifts, polished and ready to ride with a Christmas bow.

Not everyone loves building or assembling stuff, however. Here's my advice on things to do when you buy something that has to be assembled (and no, don't bring it to my house).

Pick a good place to work. Resist the temptation to open the box and dump everything out. Find an area where you can work unimpeded, and make use of a table if possible to save yourself from kneeling on the floor. If you do have to kneel on the floor (an IKEA furniture kit for example), go to the hardware store and buy a foam kneeling pad for gardeners. You won't know what you did without it.

Make sure all the parts are with the kit before you start. More than one of us has started a project only to realize not all the pieces are there. The sooner you find out, the sooner you can return to the store or call the online retailer. On some kits, a telephone number will be on the instructions and the kit manufacturer will ship parts very quickly if something is missing.

Keep the cardboard it came in to work on. Cut the box so it lays flat. This should help avoid damaging the unit and the table or floor you are working on. Be especially careful with box cutters (how do you think I found that out).

Read the directions before you do anything. This is anathema to some people. "Read the directions? Do you think I am stupid?" is usually the answer. We humans still resist looking at the directions even after disassembling and reassembling something three times because we discovered at the end that we put several parts in backwards. I have been guilty of this if it makes you feel better.

Gather the right tools before you start. After reading the directions, make sure you have all of the tools you will need at hand. There is nothing worse than holding three pieces together with one hand and realizing you don't have the Philips head screwdriver.

Be careful using power drivers. Resist the temptation to throw something together with powered bits instead of the tools they gave you in the kit unless you have a torque adjustment on the driver. Even then, be careful not to over tighten the fasteners.

Get help if you need it. Some projects will go much more smoothly with 2 people, especially furniture. This also gives you someone to complain to.

When it Pays to Not Use Shortcuts

With all of the varieties of household chores we have to do, we sometimes would like to take shortcuts and try to save some money. For example, when I was in my early 20's, my sister was upset that her kitchen faucet leaked and was old. Being young and overconfident, and wanting to impress my sister, I ran to the store and bought a new faucet assembly.

I turned off the water supply and set to disconnecting all the piping under the sink. The year was 1974 and back then we didn't have flexible water lines. The first thing I discovered was that the old supply lines weren't long enough. Oops. Back to the store.

The next thing I discovered when I returned to the kitchen was that I needed a special wrench for the fittings under the sink. Oops. Back to the store.

Hours later, after what would have taken a professional plumber about 10 minutes, I was ready to test the new faucet installation. I turned on the main water supply in the garage. I heard my sister scream from the kitchen, "It's leaking! Shut it off!" I turned off the main and ran back to the kitchen. Water was everywhere. I began sopping it up with my sister's nice towels from the guest bathroom and sighed. Oops.

Back to the store for advice. The owner of the little hardware store overheard me and came over. "It sounds like everything is hooked up correctly, I don't know why you are having leaks. Let me come take a look."

Remember, this was 1974. The owner's name was Fred, and he followed me over to the house five minutes away.

"I used to be a plumber before I started the store," he said, as he looked over my work. I was embarrassed. "Look, Lisa, you did everything right, here's your problem." He put his wrenches on one of the fittings and asked me to take them in my hands. "They are just not tight enough. Look." He helped me see how tight the connections should be. "Go turn the water on now."

I went to the garage. As I turned the main valve, I did not hear any screaming from my sis. Good sign.

I returned to the kitchen and said, "Well, I am humbled. I should have known better than to do this myself without any training."

"Don't worry about it! Said Fred. "This is how you learn. But, you can see why hiring a pro is often going to save you some time and aggravation."

We thanked Fred and tried to pay him for his time. He refused the money and smiled.

I learned a lot on that day. My sister learned a lot too – to not ask me to do any plumbing again for her!

Don't get me wrong; some of us are trained professionals, and these jobs are easy and simple, and we don't end up having to call the person we should have called in the first place. What I am saying is to carefully think out the time and cost for the project and whether its scope is within your abilities. My own stubborn nature prevented me from realizing this sooner, but now I know to think it through. I have found that much of the time the money that I spend hiring a pro, even though I COULD do the project, is actually going to save me money because I could be doing something else more productive that I already know how to do.

Is it Time to Go Solar?

In spite of the warm spell we are having at the moment, winter drives up our electric bills. Have you thought about using solar energy to offset the cost of electricity? What is really involved?

I thought about this myself as I drove past one of our many landscape solar farms. I wish they would plant trees that would really grow up around these things and make them less of an eyesore.

"Should I put solar panels on our roof?" I said to myself aloud. Fortunately, my husband was not in the car with me. I am sure he would have an opinion.

Seriously though, what is involved? Here are a few facts just to get us started. Keep in mind up front that the North Carolina tax credit has expired (35%) and the life of the 30% federal tax credit is uncertain as a new Congress takes over.

The average solar installation is about $16,000 for a 5KW system. It takes about 12 years for you to break even on your investment at current rates.

The cost of installing a solar system in your home goes down every year and the technology gets better every year, but over time the tax advantages are diminishing.

Solar systems add resale value to your home if they are maintained. You may even be able to exempt this improvement on your property taxes.

Is your roof suitable for panels? You need strength and plenty of sun for this to work. The technology has gotten so good that the orientation of your roof is less important than it used to be.

How does your homeowner's association feel about solar panels? Better to find out before the installers arrive.

Your house should already be efficient in the way it consumes electricity. There's no sense in paying for a solar system just for the heat to leak out under the front door and around the windows. You might be able to get a free energy audit from your electric provider. Give them a call.

Interconnection and net metering – how friendly is North Carolina to those of us wanting to go solar? Interconnection means how difficult is it to get the power company to accept our solar system at the meter where it interfaces with them; and net metering is where the power company agrees to credit you with power that comes back their way from the system.

North Carolina is rated "A" for interconnection, and "C" for net metering. What this means for you if you go solar is that it will be pretty easy to get the power company to hook in to your system, but they will keep more of the credits.

If you're excited about renewable energy and are serious about investing in a system, you should do some research and run the numbers. The argument for waiting is that costs are coming down and technology is getting more efficient; the argument for jumping in now is that you will be starting your journey towards the breakeven mark and you'll see some savings right away.

One caveat: lots of companies would like to sell you panels. Just get on the internet and you'll see the top 6 or 8 returns on a Google search are ads. Going solar can be complicated. Take your time and review ALL the details before jumping in.

Tips and Tricks for Hardwood Floors

We think of hardwood floors as being easy to take care of, but this may not be the case depending on your family's habits. Wood is sensitive to sunlight, moisture, and grit, and can be scratched or gouged in a moment of inattention. Here are the top five wood flooring problems and what to do about them.

Scratches. Dogs with long toenails and children with metal toys can make you wish your living room had carpet or tile. Furniture feet also cause problems. What to do: trim the dog's toenails, and place an area carpet down for children to play on. Put round felt pads on furniture legs. Rub a raw walnut into the scratch and buff with a clean cloth. You can also use a crayon of the right shade.

Dents. Solid wood and bamboo are the most susceptible, but all wood floors are easily dented by dropping heavy objects, high heels, and moving appliances. What to do: accidents happen, but lessen the chances by putting down carpet runners or smooth plywood when moving furniture and heavy household items. Think twice about lifting heavy art work, furniture, and other projects by yourself. Depending on the size of the dent, you may be able to use the walnut trick above, or dampen a cloth and use a warm iron to draw up the wood fibers.

Moisture damage. Cupping and warping can be a result of a poor installation (unregulated moisture content), wet mopping floors, and humidity in the home that is higher than 60% or lower than 30% for extended periods. What to do: never wet mop your wood floor. Damp mopping is ok as long as the mop is only slightly damp. Run air conditioning or heat as necessary to keep the humidity under control.

Shoes and vacuum cleaners. Yes, shoes and vacuum cleaners! Both do damage to wood flooring through abrasion. Vacuums with spinning brushes load fine scratches into the wood finish, and shoes track in grit. What to do: I know you won't do this, but I love the idea: take your shoes off at the door (like the Japanese do) and put on slippers. Ok, never mind. But do put down a mat at every exterior door to clear the shoes of as much dirt as possible. Regarding vacuums, you should definitely vacuum wood floors, but make sure the attachment is sliding on a soft brush or felt. Never use the power head for carpets on your wood floor. And inspect the vacuum rollers (feet) for grit before you roll it on to the floor.

Not following directions. Using wax based products, bleach, detergents, oil soaps and abrasive cleaners will damage a wood floor. The variety and composition of wood flooring make it imperative that you use the correct products for what you have. What to do: find out what type and brand of floor you have, and consult the manufacturer about how best to clean and polish it. Use products that specifically state safe for wood.

Wood floors are durable and ecologically smart, contain far less VOCs (toxic volatile organic compounds) than carpets, hold beauty and value, and never go out of style. Use these tips to keep your floors shining.

What You Should Know About the IKEA Furniture Recall

Last year the Swedish furniture maker IKEA, the largest furniture maker in the world, recalled 29 million of their MALM 3-drawer, 4-drawer, 5-drawer and 6-drawer chests.

You may have seen the video on the news of two toddlers playing on one of these bureaus and how it tipped over on to them – amazingly they were able to extricate themselves without harm, but to date seven children have been killed and dozens of others injured from climbing on IKEA chests that were never secured to the wall even though anchors were included in the kit instructions.

The IKEA recall is a perfect example of a product recall done right. First, they recalled the chests voluntarily without waiting for the Consumer Product Safety Commission to demand it; secondly, they notified as many of the customers as they could; and, in addition to sending out repair kits, they are also going to customer homes to either remove the furniture or repair it. That's an extraordinary response.

But this issue is far larger than the IKEA drawer chests that were unstable without wall anchors. Furniture tipping over is not just an IKEA problem. And tipping items are not confined to furniture; tipping dangers apply to televisions, boxes, beds, and anything else leaning against a wall or sitting atop a bookcase.

A child dies, on average, every two weeks in accidents that involve the toppling of furniture or television sets, according to the safety commission. Every year 38,000 people visit emergency rooms for injuries related to tip-over accidents. Here's what you should know to protect yourself and your family.

Every bookcase, set of drawers, or wall hanging object needs to be secured to the wall. Most items that you buy as a kit will include this in the instructions and in the package, you should find an anchor kit. Follow the directions to the letter. If you do not find an anchor kit, find a handyperson or secure the item to the wall with a bracket yourself.

Have a large TV on top of a dresser or bookcase? This is an accident waiting to happen. The backs of TVs have multiple threaded receptacles that you can use with an eye and bolt fitting to secure it to the wall or to the back of the dresser, assuming the dresser is already secured to the wall.

Objects on top of bookcases and bureaus. Look around. If a child can see something they are interested in, they will try to reach it. Secure ceramics and other objects with heavy duty Velcro or bottom brackets.

Wall objects such as mirrors and televisions should be secured with brackets designed for their weight. In fact, I'd rather buy a bracket that can handle twice the load, just to be sure. These brackets should be secured to the wall studs, not the drywall. Drywall anchors are fine for light photographs, but industrial sized and weighted items such as 65 pound TVs must have a secure installation.

Children of all ages see the home as a vast playground of toys and obstacles courses. At their young age, they are clever enough to understand that they need to climb up to reach something, but the brain hasn't developed enough for them to know that objects have mass and balance. Don't let a tip over hurt family or property: give your home a safety audit today.

Should You Build a Treehouse?

Do your kids want a treehouse? Maybe the child in you wants a treehouse? Sorry I mentioned it? The cat is out of the bag. Everyone loves a treehouse.

We live in tree fort heaven. If you have some sturdy trees behind your home and have thought about a treehouse, here are the things to consider before launching the effort.

First, some definitions. If you've been to Disney World and walked through the "Swiss Family Robinson Treehouse," then you know that this is the ultimate in tree structures. The tree is known as Disneyodendron Eximus, which means "out of the ordinary Disney tree." It is completely manmade, with concrete roots. The 1,400 limbs are constructed of steel, coated with colored cement. The entire structure weighs 150 tons and has over 1,000 square feet of living space.

The next step down is a livable house that is constructed in a tree. These tree houses are complex and have utilities. Even the smallest of houses built in a tree will require special permits and permissions if any utilities are run to it.

Several steps down in complexity is what I would call a "tree playhouse," or "tree fort." These structures are usually less than 100 square feet (that's 10' by 10'), are fairly low to the ground, and have no utilities running to them. Hey, that's the fun of a tree fort; flashlights and blankets.

If a "tree playhouse" is what you're thinking about, here are some tips. If I have you thinking about your own Swiss Family Robinson treehouse, that will have to be another column!

- Pick a simple and safe design. There is plenty of information on the internet, and even "treehouse build kits" can be found. There is also no reason why you can't just elevate a play fort next to a tree instead of putting it in the tree. This is definitely easier.
- Pick a healthy, sturdy, tree. Oak, maple, fir, and large apple trees are great candidates.
- Talk to your neighbors. Stories abound of people putting up a treehouse that is visible to the neighbors and the neighbors complain. Fortunately, here in our mountain and forest paradise, we usually have enough space to maintain privacy and not encounter this problem. But don't underestimate it if you have this situation.
- Talk to your Building Department. I asked Sam Beck in our Hayesville Building Department about treehouses and he was very helpful. It is best to err on the side of consulting the

building department on your project for engineering and safety reasons even if it is small and you're not going to be running electricity to it.

The most important consideration when building a treehouse or tree fort is safety. The child in you is going to want to climb up there with your kids. Is the ladder safe? Is the platform sturdy with good design? Is there protection with rails to prevent falls? What is on the ground underneath the fort? Consider putting a circular sand box on the ground around the tree to soften a slip or fall.

Have fun, and I hope I didn't get you in trouble.

Favorite Microwave Tricks

As I enjoyed reading Linda Brandt's column on microwave ovens, I thought back to my own experience seeing the first "radar range." It was 1979, and I was visiting one of my friends. In 1979 about 2% of the American population owned a microwave oven, versus 95% today. So I was very surprised to see my boyfriend open a small chamber in the wall next to the refrigerator and place several cold hotdogs in the opening. Intrigued, I watched him close the door and turn a dial. Three minutes later he opened the door and removed the food, steaming and smelling wonderful. I couldn't believe it.

Now, of course, the microwave oven is a wonderful addition to our kitchen toolkit. There are, as Linda points out, some great recipes designed for the device. I especially like her "coffee cup scramble" which I tried and enjoyed.

With that said, you must remember that I purposefully have developed an inability to perform anything industrious in the kitchen. In other words, I avoid cooking anything more complicated than brewing coffee. When you combine this aversion to the kitchen arts with a love of gadgets, strange things happen. In the case of the microwave oven, I delight in using it for things that are unrelated to cooking. Here are my favorites.

Fun trick – take a marshmallow and place it on a large paper plate. Heat for 20 seconds and watch it expand to 20 times its size! Don't leave it in too long or . . . you'll have a marshmallow world.

Remove chewing gum from clothes. Warm a small cup of vinegar in the microwave for 45 seconds and dab the heated liquid on the gum. It should peel off.

Warm towels – lightly dampen the towel with water and then place in a plastic bag but don't seal it. Heat for 1 minute or until it's the right temperature. Wonderful for aching joints or a soothing facial.

Wash clothes. In a hotel suite and need to launder a few items? Place socks, etc. in a bowl with soapy water and place in the microwave. Heat for five to ten minutes or until it is steaming. After letting it cool, rinse and dry.

Dye Fabric. Follow the directions to mix the dye, using a microwave safe glass bowl. Put the clothing in and stir. Heat for 4 minutes, remove and let cool. Then rinse and hope your bowl isn't also dyed.

Make a multi-color crayon by taking a bunch of old crayon sticks and placing them in a greased cupcake microwave safe pan. Heat until they flow together. Then put this in the freezer. Once cold, you will have a large weird looking crayon.

Am I serious? Yes and no. Some of the above procedures are more trouble than they are worth. Maybe I'll go back to microwave cooking.

Safety notes: Don't allow children to use the microwave unsupervised. Never put anything in the microwave that has metal in it. Use a potholder mitten or towel to take things out of the microwave that are hot. Finally, be observant. Things can go wrong fast in a microwave, like that marshmallow mess I am cleaning up right now.

Secret Hiding Places in Your Home

Ready for some intrigue? What came to mind when you read the title? Either you thought about hiding yourself or you thought about hiding valuables.

If you would like a secret hiding place from family or friends, you may need to consider new construction. The "secret room" idea isn't new, and I have built several into new construction. Secret rooms have to be well designed, though, because under scrutiny someone can discover them by analyzing the outside and then searching the inside spaces. But if you are really intent on having an entire room that's very tricky to find, it is very possible to do. If you're building a home tell your designer what you want and then call the Hidden Door Company. If you'd just like to place your jewelry in a safe place while you go on vacation, then I have some ideas for you.

There are poor places to hide your cash. Don't put valuables under your mattress - believe it or not - this is the first place burglars look. Similarly, don't place things under an obvious rock in the yard, in your sock drawer, DVD cases, behind the toilet, or on the refrigerator shelf. These are old hat.

You can try the following ideas and hope that the burglars don't read this column. You may need to be creative and get some do-it-yourself help to fabricate some of these tricks. They work because they are generally too much trouble for burglars to check out.

- Bathtub panel. Depending upon what kind of tub you have, you may be able to either cut a small access hole and cover with something that looks like it belongs – attach with Velcro – or use the space next to the motor if you have a jetted tub.
- Air vent. Buy a small fireproof box, open one of your wall vents, and place the box inside. Make sure the box does not take up more than 10% of the space or you might slow down your HVAC.

- Inside a picture. This trick works because burglars don't feel like tearing every single picture off your wall looking for things. The trick is to have multiple pictures made up as canvas prints. These prints are inexpensive and have one or more inches of space behind them. Simply cover the back with cardboard leaving a flap that can be closed with Velcro.
- Inside a hollowed out book. Another old trick that thieves will not spend time on. Unless you have zero books on your shelves, it's just too much trouble to check them all out when thieves just want to find something and leave.
- Pantry shelf containers. Take a soup can, for example, and turn it into a storage container for jewelry or cash. Make sure that it's behind the other cans or someone might get a nice surprise when they are hungry.

The best place to put your valuables is in your bank's safe deposit box. If you need fast access to some of your things, the next best place is a fire resistant safe bolted to the floor in a closet. Only the most determined thief will take the time to rip your safe out, and they will likely have trouble carrying it away unless they have an accomplice.

Now think up your own even more clever ideas. You'll be amazed at your ingenuity.

More Secret Spaces: Secret Room?

A Progress reader reminded me that you do not need to design and build a new home in order to make a secret room. Indeed, this is very true. I was thinking that most readers might not want to convert an existing room in to a "secret" space. But as I thought more about it, I decided that this might be both fun and easy to do with closet or pantry space. Local cabinet makers can look at your existing space and the entry to it and design a door that functions as a china or glass cabinet, or as a bookcase. The thickness, usually 8 or 10 inches, is visible from the secret room but sits flush with the wall on the outside. The door uses a strong piano hinge and handles substantial weight. You can also build a door that sits flush with bookcases you already have, and moves inward to reveal a doorway. If the design is clever and the matching is good, no one can tell that there is a space or room behind the cabinet. It is unlikely that a burglar would take the time to consider where a hidden door might be – they go for the obvious entries and exits.

Another reader asked if I could provide more secret space ideas. Sure!

The false bottom drawer trick. This is a fairly easy project and makes valuables highly accessible but very hard to find if you don't know about it. This is perfect for cash and jewelry. Decide what drawer you will use and measure the interior. Cut a piece of ¼ inch plywood 1/16 inch smaller than the interior. Decide how much secret space you need in the bottom and place some wood strips on the sides of the drawer at the bottom to rest the new bottom on. Glue these in. Paint the bottom to match the rest of the drawer. Place a plastic organizer into the drawer on top of the false bottom and glue it to the bottom. Now, when you need to access the space, just lift out your organizer and there are your pearls.

Hiding your front door key. Don't use the obvious places such as under the mat, above the door on a ledge, or under a rock near the door. Instead, get creative. Stand back from the door and look around. Hollow fake rocks will work if they are not near the door; put one under another rock so it's not obvious. Get a magnetic key box and look around for a good spot where it cannot be seen but it can be felt. A good trick for a near door space is an electrical junction box near ground level at the wall. Get the plastic one with a short section of plastic pipe. Cut off the end of the screw head and glue it into the bottom hole of the box. Secure the box and conduit in the ground by the door. All you need to do is push the cover to one side and there is the key.

Remember that although these tricks work and are fun, they will never replace a safe deposit box or a fire proof or fire resistant safe in your home. A fire could wipe out all of these secret places along with what you have in them.

25

Home Inspection Questions and Answers

Everything You Wanted to Know About Home Inspection But Were Afraid to Ask

A home inspection is an evaluation of the visible and accessible systems and components of a home (plumbing, heating and cooling, electrical, structure, roof, etc.) and is intended to give the client (buyer, seller, or homeowner) a better understanding of the home's general condition.

Most often it is a buyer who requests an inspection of the home he or she is serious about purchasing. A home inspection delivers data so that decisions about the purchase can be confirmed or questioned, and can uncover serious and/or expensive to repair defects that the seller/owner may not be aware of. It is not an appraisal of the property's value; nor does it address the cost of repairs. It does not guarantee that the home complies with local building codes or protect a client in the event an item inspected fails in the future.

A home inspection should not be considered a "technically exhaustive" evaluation, but rather an evaluation of the property on the day it is inspected, taking into consideration normal wear and tear for the home's age and location.

A home inspection can also include, for extra fees, Radon gas testing, water testing, energy audits, pest inspections, pool inspections, and several other specific items that may be indigenous to the region of the country where the inspection takes place.

Home inspections can be used by a seller before listing the property to see if there are any hidden problems that they are unaware of, and also by homeowners simply wishing to care for their homes, prevent surprises, and keep the home investment value as high as possible.

The important results to pay attention to in a home inspection are:

1. Major defects, such as large differential cracks in the foundation; structure out of level or plumb; decks not installed or supported properly, etc. These are items that are expensive to fix, which we classify as items requiring more than 2% of the purchase price to repair.
2. Things that could lead to major defects - a roof flashing leak that could get bigger, damaged downspouts that could cause backup and water intrusion, or a support beam that was not tied in to the structure properly.
3. Safety hazards, such as an exposed electrical wiring, lack of GFCI (Ground Fault Circuit Interrupters) in kitchens and bathrooms, lack of safety railing on decks more than 30 inches off the ground, etc.

Your inspector will advise you about what to do about these problems. He/she may recommend evaluation - and on serious issues most certainly will - by licensed or certified professionals who are specialists in the defect areas. For example, your inspector will recommend you call a licensed building or structural engineer if they find sections of the home that are out of alignment, as this could indicate a serious structural deficiency.

Home Inspections are only done by a buyer after they sign a contract, right?

This is not true. A home inspection can be used for interim inspections in new construction, as a maintenance tool by a current homeowner, a proactive technique by sellers to make their home more sellable, and by buyers wanting to determine the condition of the potential home.

Sellers, in particular, can benefit from getting a home inspection before listing the home. Here are just a few of the advantages for the seller:

1. The seller knows the home! The home inspector will be able to get answers to his/her questions on the history of any problems they find.
2. A home inspection will help the seller be more objective when it comes to setting a fair price on the home.
3. The seller can take the report and make it into a marketing piece for the home.
4. The seller will be alerted to any safety issues found in the home before they open it up for open house tours.
5. The seller can make repairs leisurely instead being in a rush after the contract is signed.

Why should I get a home inspection?

Your new home has dozens of systems and over 10,000 parts - from heating and cooling to ventilation and appliances. When these systems and appliances work together, you experience comfort, energy savings, and durability. Weak links in the system, however, can produce assorted problems leading to a loss in value and shortened component life. Would you buy a used car without a qualified mechanic looking at it? Your home is far more complicated, and to have a thorough inspection that is documented in a report arms you with substantial information on which to make decisions.

Why can't I do the inspection myself?

Most homebuyers lack the knowledge, skill, and objectivity needed to inspect a home themselves. By using the services of a professional home inspector, they gain a better understanding of the condition of the property; especially whether any items do not "function as intended" or "adversely affect the habitability of the dwelling" or "warrant further investigation" by a specialist. Remember that the home inspector is a generalist and is broadly trained in every home system.

Why can't I ask a family member who is handy or who is a contractor to inspect my new home?

Although your nephew or aunt may be very skilled, he or she is not trained or experienced in professional home inspections and lacks the specialized test equipment and knowledge required for an inspection. Home inspection training and expertise represent a distinct, licensed profession that employs rigorous standards of practice. Most contractors and other trade professionals hire a professional home inspector to inspect their own homes when they purchase a home.

What does a home inspection cost?

This is often the first question asked but the answer tells the least about the quality of the inspection. Fees are usually based on size, age and other aspects of the home. Inspection fees from a certified professional home inspector generally start under $400. An average price for a 2,000 square foot home nationally is about $375-$400. What you should pay attention to is not the fee, but the qualifications of your inspector. Are they nationally certified (passed the NHIE exam)? Are they state certified if required? How many years of experience do they have? I'd also advise against choosing an inspector with less than 3 years of experience; they will drive you crazy on the details.

How long does the inspection take?

This depends upon the size and condition of the home. You can usually figure 1.2 hours for every 1,000 square feet. For example, a 2,500 square foot house would take about 3 hours. If the company also produces the report at your home, that will take an additional 30-50 minutes.

Do all homes require a home inspection?

Yes and No. Although not required by law in most states, I think that any buyer not getting a home inspection is doing themselves a great dis-service. They may find themselves with costly and unpleasant surprises after moving into the home and suffer financial headaches that could have been avoided.

Should I be at the inspection?

It's a great idea for you be present during the inspection - whether you are buyer, seller, or homeowner. With you there, the inspector can show you any defects and explain their importance as well as point out maintenance features that will be helpful in the future. If you can't be there, it is not a large problem since the report you receive will be very detailed.

If you are not present, then you should be sure to ask your inspector to explain anything that is not clear in the report. Also read the inspection agreement carefully so you understand what is covered and what is not covered in the inspection. If there is a problem with the inspection or the report, you should raise the issues quickly by calling the inspector, usually within 24 hours.

If you want the inspector to return after the inspection to show you things, this can be arranged and is a good idea. But, you will be paying for the inspector's time on a walkthrough, usually $100 to $125.

Should the seller attend the home inspection that has been ordered by the buyer?

The seller will be welcome at the inspection (after all, it is their home), although they should understand that the inspector is working for the buyer. The conversation that the inspector has with the buyer may be upsetting to the seller if the seller was unaware of the items being pointed out, or the seller may be overly emotional about any defects. This is one reason why the seller might want to consider getting their own inspection before listing the home.

Can a house fail a home inspection?

No. A home inspection is an examination of the current condition of a home. It is not an appraisal, which determines market value, or a municipal inspection, which verifies local code compliance. A home inspector, therefore, cannot not pass or fail a house. The inspector will objectively describe the home's physical condition and indicate which items are in need of repair or replacement.

What is included in the inspection?

The following list is not exhaustive. Not all of these may be in the inspection you get, but the inspector will be following a standardized checklist for the home:

Site drainage and grading
Driveway
Entry Steps, handrails
Decks
Masonry
Landscape (as it relates to the home)
Retaining walls
Roofing, flashings, chimneys, and attic
Eaves, soffits, and fascias
Walls, doors, windows, patios, walkways
Foundation, basement, and crawlspaces
Garage, garage walls, floor, and door operation
Kitchen appliances (dishwasher, range/oven/cooktop/hoods, microwave, disposal, trash compactor)
Laundry appliances (washer and dryer)
Ceilings, walls, floors
Kitchen counters, floors, and cabinets
Windows and window gaskets
Interior doors and hardware
Plumbing systems and fixtures
Electrical system, panels, entrance conductors

Electrical grounding, GFCI, outlets
Smoke (fire) detectors
Ventilation systems and Insulation
Heating equipment and controls
Ducts and distribution systems
Fireplaces
Air Conditioning and controls
Heat Pumps and controls
Safety items such as means of egress, TPRV valves, railings, etc.

Other items that are not a part of the standard inspection can often be added for an additional fee:

Radon Gas Test (if your home falls in a high radon area)
Water Quality Test
Termite Inspection (usually performed by a separate company)
Gas Line Leak Test (usually performed by the gas company)
Sprinkler System Test
Swimming Pool and Spa Inspection
Mold Screening (sometimes performed by a separate company)
Septic System Inspection (usually performed by a separate company)
Alarm System (usually performed by a separate company)

Your inspector will work with other companies on some of these specialized inspections. You should ask about the arrangement and what extra fees might be involved.

What is not included in the inspection?

Most people assume that everything is inspected in depth on inspection day. This misunderstanding has caused many a homebuyer to be upset with their inspector.

If you hired someone with licenses for heating and cooling, electrical, plumbing, engineering, etc. to inspect your house, it would take about 16 hours and cost you about $3,000. It is much more practical to hire a professional inspector who has generalist knowledge of home systems, knows what to look for, and can recommend further inspection by a specialist if needed.

Your inspector is also following very specific guidelines as he/she inspects your home. These are either national guidelines (ASHI - American Society of Home Inspectors, InterNACHI - International Association of Certified Home Inspectors) or state guidelines.

These guidelines are carefully written to protect both your home and the inspector. Here are some examples: We are directed to not turn systems on if they were off at the time of the inspection (safety reasons); we are not allowed to move furniture (might harm something); not allowed to turn on water if it is off (possible flooding), and not allowed to break through a sealed attic hatch (possible damage).

The downside of this practice is that by not operating a control, by not seeing under the furniture, and not getting into the attic or crawlspace, we might miss identifying a problem. However, put into perspective, the chances of missing something serious because of this is quite low, and the guideline as it relates to safety and not harming anything in the home is a good one.

There are other items that 95% of inspectors consider outside a normal inspection, and these include inspecting most things that are not bolted down (installed in the home) such as electronics, low voltage lighting, space heaters, portable air conditioners, or specialized systems such as water purifiers, alarm systems, etc.

What if there are things you can't inspect (like snow on the roof)?
It just so happens that some days the weather elements interfere with a full home inspection. If there is snow on the roof the inspector will tell you they were unable to inspect it. They will be looking at the eves and the attic, and any other areas where they can get an idea of condition, but the inspector will write in the report that he/she could not inspect the roof. It may be impractical to return another day once the snow melts,

because inspectors have full schedules. However, you can usually pay an inspector a small fee to return and inspect the one or two items they were unable to inspect when they were there the first time. This is just the way things go. If you ask the inspector for a re-inspection, they will usually inspect the items then at no extra charge (beyond the re-inspection fee).

Will the inspector walk on the roof?

The inspector will walk on the roof if it is safe, accessible, and strong enough so that there is no damage done to it by walking on it. Some roofs - such as slate and tile, should not be walked on. Sometimes because of poor weather conditions, extremely steep roofs, or very high roofs, the inspector will not be able to walk the roof. The inspector will try to get up to the edge though, and will also use binoculars where accessibility is a problem. They will also examine the roof from the upper windows if that is possible. There is a lot the inspector can determine from a visual examination from a ladder and from the ground, and they will be able to tell a lot more from inside the attic about the condition of the roof as well.

Should I have my house tested for Radon? What exactly is Radon?

In many areas of the country, the answer is a definite yes. You can ask your real estate agent about this or go on to the internet for a radon map of the country. Radon is a colorless, odorless, tasteless radioactive gas that's formed during the natural breakdown of uranium in soil, rock, and water. Radon exits the ground and can seep into your home through cracks and holes in the foundation. Radon gas can also contaminate well water.

Health officials have determined that radon gas is a serious carcinogen that can cause lung cancer, second only to cigarette smoking. The only way to find out if your house contains radon gas is to perform a radon measurement test, which your home inspector can do. Make sure the person conducting your test has been trained to The National Environmental Health Association (NEHA) or The National Radon Safety Board (NRSB) standards.

What about a newly constructed home? Does it need a home inspection?

Yes. In fact, inspectors find far more problems, some quite serious, in newly constructed homes than in homes that have been lived in for years. This is not due to your builder's negligence - he/she has done the best job they could with subcontractors and planning - it's just that there are so many systems in a home, that it is close to impossible to inspect everything, and correct it before the Certificate of Occupancy is issued.

Then, for some reason, the subcontractors no longer want to work on the home, and final jobs and details are missed. We recommend getting several professional home inspections near the completion stages of the home to discover everything that should be corrected.

If the house is still new but sitting for a while before sale, it's even more important to get a home inspection. I have seen water lines not hooked up, plumbing lines not hooked up, sewer lines not hooked up, vents not hooked up, and a variety of other serious but easily correctable problems.

I am having a home built. The builder assures me he will inspect everything. Should I have an independent inspector make periodic inspections?

Absolutely yes. No matter how good your builder is, he/she WILL miss things. They are so concerned with the house, they get so close to their work, as do the subcontractors, that important items can, and will be, overlooked. Have a professional inspector make at least 4-6 interim inspections. They will be worth their weight in gold.

What is the Pre-Inspection Agreement?

Most service professionals have a service agreement, and home inspection is no different. In fact, there is enough confusion about what a home inspection should deliver that the agreement is even more important. Some homeowners who get a home inspection expect everything in the home to be perfect after the repairs. This is not the case. Imagine getting a call from a homeowner a year later who says the toilet is not flushing - remember that the inspection is a moment in time snapshot.

In the inspection agreement, the inspector is clear about what the inspection delivers and the things that are not covered, as well as what you should do if you are not pleased with the services. By reviewing this before-hand you will understand much more about the inspection and be happier with the results.

A home inspection does not guard against future problems, nor does it guarantee that all problems will be found.

What kind of report will I get following the inspection?

There are as many versions of a "report" as there are inspection companies. Guidelines dictate that the inspector deliver a written report to the client. This can range from a handwritten checklist that has multiple press copies without pictures to a computer generated and professionally produced report with digital pictures 35 pages long and can be converted to Adobe PDF for storage and emailing.

Check with your inspector about the report he or she uses. I recommend the computer generated report, since the checklist is more detailed and easier for the homeowner/buyer/seller to detail out the issues with photographs. In this modern age, the reports must be web accessible and emailable to match the technologies most of us are using.

There are some great things you can use the report for in addition to the wealth of information it simply gives you on your new home:

- Use the report as a checklist and guide for the contractor to make repairs and improvements or get estimates and quotes from more than one contractor.
- Use the report as a budgeting tool using the inspector's recommendations and the remaining expected life of components to keep the property in top shape.
- If you are a seller, use the report to make repairs and improvements, raising the value of the home and impressing the buyers. Then have a re-inspection and use this second report as a marketing tool for prospective buyers.

- Use the report as a "punch list" on a re-inspection and as a baseline for ongoing maintenance.

Will the report be emailable or available as an Adobe PDF file?
Yes. As discussed in the last question, you will probably want your inspector to be using the latest reporting technology.

What if I think the inspector missed something?
Inspectors are human, and yes, they do miss items. However, they routinely use advanced tools and techniques to reduce the possibility that they will miss something. This includes very detailed checklists, reference manuals, computer based lists, and a methodical always-done-the-same-way of physically moving around your home. That is one of the reasons that an inspector can miss an item when they get interrupted. The inspector will have a set way of resuming the inspection if this happens. If, in the end, something IS missed, call the inspector and discuss it. It may warrant the inspector returning to view something that you found. Remember, the inspector is doing the very best job they know how to do, and probably did not miss the item because they were lax in their technique or did not care. When I was an inspector, I always returned to look at something at no charge if the client asked.

What if the inspector tells me I should have a professional engineer or a licensed plumber or other professional contractor in to look at something they found? Isn't this "passing the buck"?
You may be disappointed that further investigation is required, but your inspector is doing exactly what they should be doing. The purpose of the inspection is to discover defects that affect your safety and the functioning of the home; the inspector is a generalist, not a specialist.

The inspection code of ethics as well as national and state guidelines dictate that only contractors that are licensed in their specialty field should work on these systems and areas. When they tell you that a specialist is needed, there may be a bigger, more critical issue that you need to know about.

If you move into the home without getting these areas checked by a qualified specialist, you could be in for some nasty and expensive surprises. The inspector does not want to cause you any more expense or worry, so when they do recommend further evaluation they are being serious about protecting you and your investment.

Will the inspector provide a warranty on the inspected items?

Most inspectors do not give the homeowner a warranty on inspected items. Remember, a home inspection is a visual examination on a certain day, and the inspector cannot predict what issues could arise over time after the inspection. However, some inspectors are now including a warranty from the largest home warranty company in America - American Home Warranty Corporation, as well as others, on the inspected items for 60 or 90 days. This is a very good deal, and the agreement can be extended after the initial period for a relatively small amount of money.

Do most inspection companies offer money back guarantees?

Most inspection companies do not offer a satisfaction guarantee nor do they mention it in their advertising. It's always a good thing if you can get extra services for no additional cost from your inspection company, and of course a satisfaction guarantee is an indication of superior customer service. You usually have to call your inspection company right after the inspection and viewing of the report to tell them you are not satisfied.

If you are not happy with the services, you should talk to your inspector first and let him/her correct the issue(s) you are unhappy with first.

When I ran my own inspection company, I did offer a 100% money back guarantee to customers, and I had less than a half percent of customers ask for it. The few that did ask had misunderstood what I was inspecting (thought the inspection covered termites). That was my error in not explaining it well enough.

What if my report comes back with nothing really defective in the home? Should I ask for my money back?
No, don't ask for your money back - you just received great news. Now you can complete your home purchase with peace of mind about the condition of the property and all its equipment and systems.

You will have valuable information about your new home from the inspector's report, and will want to keep that information for future reference. Most importantly, you can feel assured that you are making a well-informed purchase decision.

What if the inspection reveals serious defects?
If the inspection reveals serious defects in the home (I define a serious defect as something that will cost more than 2% of the purchase price to fix), then pat yourself on the back for getting an inspection. You just saved yourself a ton of money. Of course it is disappointing, even heart wrenching, to find out that your well researched house is now a problem house, but you now know the facts and can either negotiate with the seller, or move on. You may want the home so much that it will be worth it to negotiate the price and then perform the repairs. Imagine, though, if you had not gotten the inspection - you would have had some very unpleasant surprises.

Can I ask my home inspector to perform the repairs?
You can, but if your inspector is ethical, he/she will refuse, and correctly so; it is a conflict of interest for the person who inspected your home to also repair it. Inspectors are specifically barred from this practice by licensing authorities, and it's a good rule - an inspector must remain completely impartial when he or she inspects your home.

This is one reason you should have a professional home inspector inspect your home and not a contractor - the contractor will want the repair work and you are likely to not have an objective inspection from this person even though they mean well and are technically competent.

Does the Seller have to make the repairs?

The inspection report results do not place an obligation on the seller to repair everything mentioned in the report. Once the home condition is known, the buyer and the seller should sit down and discuss what is in the report. The report will be clear about what is a repair and what is a discretionary improvement. This area should be clearly negotiated between the parties.

It's important to know that the inspector must stay out of this discussion because it is outside of their scope of work.

After the home inspection and consulting with the seller on the repairs, can I re-employ the inspector to come re-inspect the home to make sure everything got fixed?

You certainly can, and it's a really good idea. For a small fee the inspector will return to determine if the repairs were completed, and if they were completed correctly.

What if I find problems after I move into my new home?

A home inspection is not a guarantee that problems won't develop after you move in. However, if you believe that a problem was visible at the time of the inspection and should have been mentioned in the report, your first step should be to call the inspector. He or she will be fine with this, and does want you to call if you think there is a problem. If the issue is not resolved with a phone call, they will come to your home to look at it and they should not charge you for this. They will want you to be satisfied and will do everything they can to achieve this goal.

One way to protect yourself between the inspection and the move-in is to conduct a final walkthrough on closing day and use both the inspection report AND a Walkthrough Checklist to make sure everything is as it should be.

INDEX

ABOUT THE AUTHOR

Lisa grew up taking things apart and playing with the boys. After graduating from college, she started a bicycle shop, attended night school for an engineering degree, and took on odd jobs in residential and commercial construction.

Lisa dreamed about flying airplanes. In the mid-nineties, she bought the parts to build a kit airplane in her garage. Twenty months later she made a dream-come-true solo round trip in her homebuilt craft, called a Pulsar XP, from south Florida to Bar Harbor, Maine, to visit her family.

Lisa met her antique aircraft restorer husband, Jerry Stadtmiller, in 1999 as she was beginning to build her second airplane. They married in 2003, and several years later moved to the mountains of Hayesville, North Carolina.

In 2006, Lisa started Your Achievement Coach, a coaching and teaching practice that focused on delivering planning skills to people wanting to reach their goals. Lisa believed that whether it was starting a business, writing a book, or building an airplane, developing the skills of disciplined planning and goal setting could get you there. Lisa said to her clients, "Shine the light of possibility on your dream, and it will leap to action."

Switching gears again in North Carolina, Lisa founded and ran Your Inspection Expert, a residential inspection company, from 2008 to 2011. Experience gleaned from hundreds of residential inspections form the basis for the advice in this book.

Lisa has worked for 3 major U.S. corporations at the executive level, most notably as Chief Training Officer for Tyco Fire and Security Services in Boca Raton, Florida, in the early 2000s. She was certified as an ASQ Six Sigma Black Belt and Quality Engineer; she holds a 50-ton coast guard captain's license, a private pilot license, an FAA airframe and powerplant license (A&P), and FAA Light Sport Repair Certificate Instructor certification.

In 2001 Lisa earned the EAA (Experimental Aircraft Association) Technical Counselor designation and a few years later, Flight Advisor status. On February 5th, 2008, Lisa became the first civilian female AB-DAR (Amateur Built Designated Airworthiness Representative) for the FAA. In these roles Lisa volunteers her time to help experimental aircraft builders become technically proficient and make their airplane as safe as possible. She also helps them determine the testing program for their aircraft and understand the critical factors required for the first flight and early flying hours.

Lisa holds a Doctor of Science, a Bachelor of Arts in English and Philosophy, a Masters in Business, and an Associate of Science in Engineering.

Lisa was a United States Coast Guard volunteer in the 1980s and 90s, worked with Junior Achievement and the Flying Start program, and continues to volunteer for the Experimental Aircraft Association.

People have asked Lisa whether being a woman in male dominated professions has made her journey more difficult. Lisa answered, "No, not at all. In fact, men have encouraged me and helped me to be successful. If anything, being unconventional has made the path easier. Appreciating the contributions of both men and women in unusual endeavors brings balance."

Made in the USA
Middletown, DE
08 December 2018